Hyaluronic Acid

Hyaluronic Acid

Preparation, Properties, Application
in Biology and Medicine

MIKHAIL A. SELYANIN, PETR YA. BOYKOV
AND VLADIMIR N. KHABAROV

Martinex International Research Center, Moscow, Russia

Translated from the Russian version by Scientific Editor Felix Polyak

WILEY

This edition first published 2015
© 2015 John Wiley & Sons, Ltd.

Registered Office
John Wiley & Sons, Ltd., The Atrium, Southern Gate, Chichester, West Sussex, PO19 8SQ, United Kingdom

For details of our global editorial offices, for customer services and for information about how to apply for permission to reuse the copyright material in this book please see our website at www.wiley.com.

Library of Congress Cataloging-in-Publication Data

Khabarov, V. N. (Vladimir N.)
   Hyaluronic acid : preparation, properties, application in biology and medicine / Mikhail A. Selyanin, Petr Ya. Boykov and Vladimir N. Khabarov ; translated from the Russian version by scientific editor Felix Polyak.
      pages   cm
   Includes bibliographical references and index.
   ISBN 978-1-118-63379-3 (cloth)
1. Hyaluronic acid.   2. Organic acids.   I. Boykov, P. Ya.   II. Selyanin, M. A. (Michael A.)   III. Title.
   QP702.H8H9313 2015
   612.015782–dc23
                                        2014038400

A catalogue record for this book is available from the British Library.

ISBN: 978-1-118-63379-3 (HB)

Set in 10/12pt Times by SPi Publisher Services, Pondicherry, India
Printed and bound in Singapore by Markono Print Media Pte Ltd

1   2015

This book is dedicated to our daughters Daria, Maria and Vasilisa

Nature is not as you imagine her:
She's not a mold, nor yet a soulless mask-
She is made up of soul and freedom
She is made up of love and speech …

F.I. Tyutchev (Russian Poet)

Contents

Foreword

In March 2012, I came across a book about hyaluronic acid, which had recently been published in Russian[1]. I read it with growing interest. I found it extremely educational and insightful, especially the chapters about biology and medical applications of hyaluronic acid (HA). Surprisingly, when I attempted to find other recently published manuscripts, my searches returned no results. I use the word 'surprisingly', because there is enormous interest in this compound and an increasing, even exponential growth in the application and use of HA-based products in the cosmetic industry as well as in medicine. There is no doubt that we will see even an broader application of HA to these areas in the near future.

I found this book very thorough in its description and analysis of HA and very well written. It covered all aspects of HA; its history and discovery, its biological role and functions in the body, production and purification, structure and rheological properties, chemical modifications, application in medicine, cosmetology and aesthetic medicine. It should be taken into consideration that the manuscript was written several years ago, so the chapter about HA production could be seen as outdated. Nevertheless, we decided to leave it intact in this version, because it contains important information about HA purification for industrial needs. Additionally, it serves to allow the authors to present their significant achievement with regard to the chemical modification of HA, an accomplishment that has opened the doors to many new products.

As I mentioned, the book was published in Russian, which is why so many potential readers were not able to read it. It did not take much time for me to decide to translate it and try to find a way for English-speaking readers to enjoy the book. I would like to express enormous appreciation to the publishing house John Wiley & Sons, Ltd, for agreeing to publish it.

It is important to mention that the Russian manuscript, in addition to its great scientific value, is written with incredibly beautiful literary flair. I tried with great effort to maintain the original style and to give the English readership the opportunity to enjoy it as much as I did. It took more than a year to translate and edit the book and now it is ready to meet the readers. I would like to express my warmest appreciation to everybody who has helped me to prepare the English manuscript, especially the editors from John Wiley & Sons, who were very supportive in every step during the preparation of the manuscript.

As the authors mentioned in their Introduction, we appreciate each and every critical comment, which we will take on board in our efforts to improve upon this first version.

Felix Polyak, PhD

---

[1]The original Russian language version was published in 2012 by Prakticheskaya Meditsina in Moscow: V.N. Khabarov, P.Ya. Boykov, M.A. Selyanin, *Hyaluronic acid*. Moscow: Prakticheskaya meditsina, 2012 (in Russian).

Introduction

'Biopolymers' is the common (or general) name for biological macromolecules, which include proteins, nucleic acids and polysaccharides. The existence of the main processes of life is determined by interaction of these substances. Hyaluronic acid possesses one of the leading places among them; it is not accidentally called 'a wonderful evolution invention'. It would not be easy to come across a monograph dedicated to a single particular compound among scientific publications.

Hyaluronic acid is polysaccharide that was discovered 80 years ago. Even today, it continues to surprise researchers with its unique properties in various areas of chemistry, biology and medicine.

For example, the medical aspects include the study of the hyaluronan role in fertilization, embryogenesis, development of immune responses, wound healing in infectious and oncological diseases, in ageing processes and in aesthetic medicine problem solutions. Hyaluronan is involved in numerous biological processes. It is present in practically all body tissues, where it plays its role in the cell activity regulation or, on the other hand, slows down cell division, migration and is involved in the reorganization of chromatin structure and gene switch.

The natural question is, how did such a variety of physiological and functional properties of this polysaccharide originated, despite its quite simple chemical structure? To clarify this issue even partially, the authors attempted to link the biological functions of hyaluronic acid with the general principles of the structure and physicochemical properties, using a multidisciplinary approach.

The manuscript consists of six chapters. Many issues are discussed with the different levels of complexity and with different degrees of completeness. Every chapter has its own bibliography.

The first chapter describes the history of hyaluronic acid discovery, the main milestones of its research and practical application. The largest chapter, Chapter 2, is dedicated to the biological role of the hyaluronic acid in nature, in particular in the human body. The chapter starts from the phylogenesis of hyaluronic acid, then describes hyaluronan functions in human ontogenesis and especially the role that hyaluronan plays in the extracellular matrix of the different tissues.

The problems of production and purification of hyaluronic acid are discussed in detail in Chapter 3. The comparable data of the commercial products are presented in Chapter 3 as well. The structure and rheological properties are discussed in Chapter 4. The discussion is united by a general idea about cooperative conformational transformations as the main phenomena that manifest bio-specificity of hyaluronic acid as well its unique effect of the relationships of biological properties from molecular mass. The physical and chemical methods of hyaluronan structural modification are presented in Chapter 5. It presents reactions of the macro-molecule cross-linking with bi-functional reagents as well as highly promising method of hyaluronic acid solid state modification. This method, which applies both high pressure and sheer deformation, seems to be very promising both scientifically and practically. The application of the solid state approach allows production of the unique medical products, even now. These preparations, which belong to the class of drug-less (or non-drug) macro-molecular therapeutic products, are prepared from hyaluronan, modified with different vitamins, amino acids and oligopeptides. The final Chapter 6 provides overview of the medical application of hyaluronic acid. In addition to broad application in aesthetic medicine and mesotherapy, hyaluronan has become very popular in such areas as ophthalmology, arthrology, wound healing, immunology, oncology and so on. With better understanding of the biological role of hyaluronan and development of the chemical and physical methods of its derivatization, we can expect increasing role of hyaluronic acid and its products in medicine.

The readers will find that the central part of the book is dedicated to the biological role and medical application of that polysaccharide. The comprehensive scientific material allowed the authors have a new look at the extremely important place of hyaluronic acid in the multi-face world of bio-molecules.

The book is intended for the broad range of the readers – students, professors and researchers, who are interested in the chemistry of biopolymers, molecular biology and medicine.

The authors are extremely grateful for the valuable advice that they received from Professor A. Shilov, Professor N. Milanov, Professor A. Zeleneckiy, Professor A. Shehter, Dr A. Safoyan, Professor I. Samoilenko and Dr V. Volkov. These sincere words are addressed to everybody who helped to prepare this manuscript.

The authors admit that the book is obviously not free from drawbacks, caused by the broad amount of factual material and are grateful in advance for critical comments.

1

The History of Hyaluronic Acid Discovery, Foundational Research and Initial Use

1.1 Discovery

In 1934, Karl Meyer and John Palmer wrote in the *Journal of Biological Chemistry* about an unusual polysaccharide with an extremely high molecular weight isolated from the vitreous of bovine eyes [1]. Being the first to mention it, they gave the new substance the name *hyaluronic acid* (HA, the modern name 'hyaluronan') derived from 'hyaloid' (glassy glass-like in appearance) and 'uronic acid'. While Meyer and Palmer are generally considered to have discovered hyaluronic acid, it is fair to mention that as far back as 1918 Levene and Lopez-Suarez had isolated a new polysaccharide from the vitreous body and cord blood that they called 'mucoitin-sulfuric acid' [2]. It consisted of glucosamine, glucuronic acid and a small amount of sulfate ions. It is now clear this substance was actually hyaluronic acid extracted together with a mixture of sulfated glycosaminoglycans.

At the time of the discovery of hyaluronan, the polysaccharides, which represent the major part of the organic material on our planet, were already quite well known. A number of so-called mucopolysaccharides, currently known as glycosaminoglycans, had already been discovered. Hyaluronic acid is known to belong to this class as well. Mucopolysaccharides were isolated from mucus, to which they give viscous lubricating properties. These properties, in turn, are related to glycosaminoglycan's ability to bind to a significant amount of water.

Hyaluronic Acid: Preparation, Properties, Application in Biology and Medicine, First Edition.
Mikhail A. Selyanin, Petr Ya. Boykov and Vladimir N. Khabarov.
© 2015 John Wiley & Sons, Ltd. Published 2015 by John Wiley & Sons, Ltd.

1.2   Foundational Research

Soon after the original work was published, unique properties of the new biopolymer were discovered, which proved it different from other similar glycosaminoglycans, According to Meyer and Palmer, the isolated polysaccharide contained uronic acids and amino sugars, as well as pentose, and was not sulfated [1]. They also decided that the molecular mass of the repeatable unit is approximately 450 Da. It was later proved that HA in fact does not contain sulfate groups or pentose. It was also established that the molecular mass of the repeatable disaccharide residue is 397 Da.

Over the next 10 years, Meyer and other authors isolated hyaluronan from various animal organs. For example, the polysaccharide was found in joint fluid, the umbilical cord and recently it has become possible to extract HA from almost all vertebrate tissues. In 1937, F. Kendall isolated hyaluronan from the capsules of *streptococci* groups A and C. This work had great scientific and practical importance, as today *streptococci* groups are the most economical and reliable source for the industrial production of hyaluronic acid [3].

In 1928, F. Duran-Reynals found a certain biologically active compound in rabbit testicles that lead to an extremely important discovery in the chemistry and biology of hyaluronic acid. When the compound was injected with black indian ink subcutaneously, the authors observed extremely fast distribution of the black colour through connective tissue [4]. Similar properties were found for the extracts from semen, leeches, bee sting and snake venom. Further studies confirmed that the observed increasing permeability of connective tissue was mainly caused by the depolymerization of its basic substance, hyaluronic acid. It was thus determined the extract contains a specific enzyme that was given the name 'hyaluronidase'. The biological material that contains hyaluronidase was recently called the Duran-Reynals *spreading factor.*

The discovery of enzymes that could selectively break down hyaluronan opened the door for the establishment of the polysaccharide molecule's chemical structure. In those days, a powerful tool for analysing the structure of polysaccharides such as nuclear magnetic resonance spectroscopy NMR was not known. At the present time, NMR makes it possible to determine the monosaccharide biopolymer residue's composition, centres for substitution reactions, sequence and three-dimensional structure.

In 1943 E.A. Balazs and L. Piller published a paper in which they described a study of role of hyaluronan in dog knee joints. They found that the intercellular substance of connective tissue of the synovium contains sufficient viscous mucin that can replace the mucin removed from the knee [5]. These observations literally opened the door to further studies on the role of hyaluronan in normal and traumatic joints. In 1949, C. Ragan and K. Mayer published a very important paper in which they described the observation of hyaluronan in rheumatoid synovial fluid. This was the first study in which normal and pathological synovial fluids were compared by determination of the concentration and viscosity of hyaluronan [6].

In the short period between 1948 and 1951, several chemists initiated research to elucidate the structure of hyaluronic acid. In 1948 A. Dorfman published the first results of a kinetics of fermentative hydrolysis of hyaluronan [7]. Three years later in 1951, A.G. Ogston and J.E. Stanier published the first significant data about the structure of the HA macromolecule in aqueous solution. They found that the relationship between viscosity and velocity gradients increased with higher concentrations of the polysaccharide. [8]. It was found that this phenomenon is due to the interlacement of the neighbouring molecules, not individual macromolecule asymmetry. In 1955 an irregular helical configuration of hyaluronan was confirmed by measuring light scattering [9].

Several major research directions on hyaluronic acid were identified in the first half of the twentieth century. Lately, they have developed into independent branches within different fields of science including polymer chemistry, radiochemistry, biochemistry, molecular biology, medicine and glycobiology. The latter term was accepted in 1988 to describe a branch of science that combines a traditional biochemistry of hydrocarbons with a modern understanding of the role of complex sugars in cell and molecular biology.

Causing particular curiosity and scientific wonderment for researchers was the different observed viscosities of the hyaluronan solutions in presence of the different inorganic salts. The largest viscosity was observed for the solution in distilled water. It was proposed that the viscosity could be related to pH values and solution ionic strength. This phenomenon has become common knowledge but was initially described by R. Fuoss only for solutions of the synthetic polyelectrolytes [10].

Fundamental research on the physico-chemical properties of HA is considered to have begun in 1951 with the publication of E.A. Balazs's article [11]. One of the first attempts to sterilize HA by UV light led to a complete loss of the solution viscosity. A similar result was obtained by A. Caputo in 1957 by X-ray exposure of the hyaluronan solution [12]. Later, it was found that when exposed to gamma radiation or electron beams, even at low initial levels of absorbed dose of ionizing radiation, HA degrades completely. The processes of polysaccharide radiolysis, which are associated with polymer degradation and involve free radicals, are now intensively studied in the radiochemistry of biomolecules.

Unlike sulfated polysaccharides, some of the initial proof of HA's ability to interact with living cells came with the observation that hyaluronan accelerates cell growth. It has also been observed that hyaluronan initiates some cell aggregation. This was the first indication of a unique binding of the polysaccharide to the cell surface. Currently, several receptor proteins that bind to the surface of the HA cytoplasmic membrane have been isolated, including high-affinity receptor CD44 and receptor RHAMM (receptor for hyaluronan-mediated motility).

The receptor for HA endocytosis had been found on the membrane of endothelial cells of the liver sinuses and fundamentally differs from other hyaluronan-binding proteins (see [13] and references therein).

These early studies accomplished much in a short period of time, notably the establishment of the structure and monomeric composition of the macromolecule. In 1954, Meyer published an article in *Nature* that presented the result of a study on the decomposition products of HA [14]. The article included the structural formula of the disaccharide, which is the product of HA cleavage by streptococcus hyaluronidase (Figure 1.1).

1.3   Initial Medical Applications

During the second half of the twentieth century, HA was discovered in different tissues and liquids of vertebrae animals as well as humans. It was also found to have clinical applications, mostly for eye surgery, treatment of joint diseases and aesthetic medicine. The first actual use

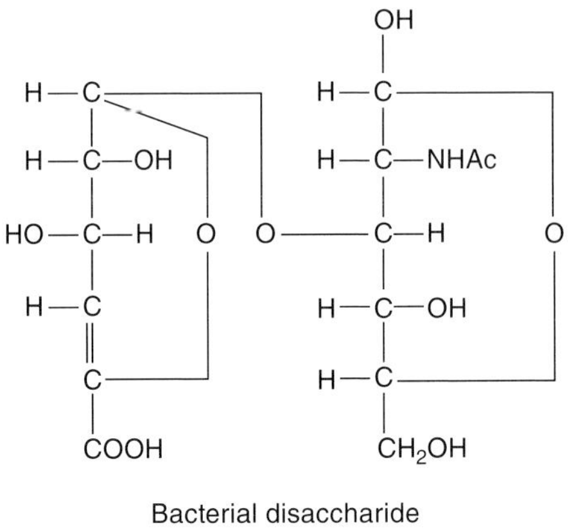

Bacterial disaccharide

Figure 1.1 *Structure of 4,5-unsaturated disaccharide, obtained by HA cleavage by bacterial hyaluronidase*

of HA in medicinal practise didn't actually occur until 1943 during the Second World War. N.F. Gamaleya (Н.Ф. Гамалея) created complex bandages in order to treat the frostbitten soldiers in the military field hospital no. 1321. The main component of the bandage was an extract from the umbilical cord, which he called a 'factor of regeneration'. The method was later approved by the USSR Ministry of Health and the drug received the name 'Regenerator'. It is apparent that HA was a major contributor towards the positive effect of the treatment, given that the human umbilical cord contains a significant amount of HA. In fact, at this time the umbilical cord was considered to be one of the most important industrial sources of HA alongside other biological materials.

Several practical ventures that explored HA's medical applications followed. In the 1950s, E.A. Balazs initiated experiments with HA to investigate its potential as a prosthesis for the treatment of retinal detachment. In 1969 it was reported that HA was used in order to prevent postoperative soldering. In 1970, hyaluronan was first injected into the joints of racehorses that suffered from arthritis with a clear and positive outcome observed. A few years later, R. Miller started to use HA in implanted intraocular lenses [15]. Since these ground breaking cases, hyaluronan has become one of the most important components in ophthalmology (see the review about hyaluronan application in ophthalmology and the references therein [16]) and has found extremely wide application in aesthetic medicine. Today, HA-contained products are the 'gold standard' for injectable cosmetics.

1.4 Sources of Hyaluronan

Due to its increasing applications, the demand for hyaluronan has grown from the moment of its discovery up until now. As the aforementioned branches of medicine and cosmetology widen, the role of HA is being reconsidered from the passive structural matrix of connective tissue to the understanding of the primary role of that macromolecule in many important physiological processes. These processes include cell communication, migration, differentiation, process regulation in the extracellular matrix and activation of cell structure metabolism.

At the present time, it is known that HA is not an inactive macromolecule of connective tissue, but a metabolically highly active biopolymer. Its half-life in the joints is 1–30 weeks, up to 1–2 days in the epidermis and derma and only 2–5 minutes in the bloodstream. In other words, during one day, approximately 5 g of dry HA can be synthesized and cleaved in the body of an adult 70-kg man, one-third of the whole amount of HA in the body [17].

The HA polysaccharide chain undergoes degradation by endoglucanase (hyaluronidase) and exoglucanase (beta-glucouronidase and beta-N-acetyl hexosaminidase). The testicle hyaluronidases decompose polysaccharides with a hydrolysis of the glycoside bond to tetra-, octa- and other saccharides.

It was pointed out that HA was first discovered as an animal polysaccharide, but soon thereafter it was found that the biopolymer also exists among bacteria. In 1937, Kendall, Heidelberger and Dawson reported about extraction of a polysaccharide from the cultural liquid of the haemolytic streptococcus that was precipitated with acetic acid and ethanol [3]. The authors proposed that the isolated biopolymer is identical to hyaluronic acid, a hypothesis that was later confirmed. It was eventually found that the mammalian glycosaminoglycan exists among several groups of the streptococcus, many of which are pathogenic for humans and animals.

HA was initially produced by extraction from the animal material. Because of the growing demand of HA, however, scientists started looking for new methods for its production, methods that were preferably microbiological. A study of the bacterial synthesis of hyaluronan on the free cells level started when UDP-glucose, UDP-N-acetyl glucosamine and UDPP-glucuronic acid were obtained from streptococcus extracts.

In 1953 Roseman and coworkers published an article in which they describe the precipitation of HA from the cultural liquid (CL) of Group A streptococcus [18]. They reported the yield 200–300 mg from 4 l of CL. Later, Warren and Gray found the semi-synthetic media for the cultivation of the HA producers [19].

A significant number of patents have been filed on the production and clinical application of HA. The production of HA by cultivation of the single strain *Streptococcus equi* was described in more than 20 patents from 1985 to 2002. Despite these efforts, however, the problem of HA supply failing to meet worldwide demand remains. The price for highly purified pharmaceutical grade HA has reduced dramatically during last few years, but the current minimum price is still at the level of $10 000/kg. To compare, the price of the polysaccharide xanthan produced from *Xanthomonas campestris* is $11/kg. The enzymes responsible for the metabolism of hyaluronan have been studied since 1959, when the hyaluronate synthase (HAS) from *Streptococcus pyogenes* was first described [20]. Many researchers tried to identify, solubilize and isolate a pure active enzyme that would be able to synthesize HA both in streptococcus and in eukaryotic cells or to clone HAS gene in *E. coli*. Unfortunately, after several decades of persistence, their efforts were not successful.

In 1993 a group of US scientists reported about isolation of HAS, its operon and cloning of hyaluronate synthase gene into *E. coli* [21]. Their findings were recently found to be scientific error. Van de Rijn and Drake isolated three streptococcus membrane proteins with molecular masses of 42, 33 and 27 kDa and proposed that the protein with mass 33 kDa is indeed hyaluronate synthase [22]. Using electrophoresis, other studies have found that streptococcus hyaluronate synthase has a molecular mass 42 kDa, proving this conclusion incorrect.

Soon thereafter, a breakthrough in a study of the mechanism of HA synthesis and regulation occurred. Almost simultaneously, DeAngelis and co-workers reported that the operon of the

HA synthesis was found, isolated, characterized and cloned [23]. It was the first successfully cloned hyaluronate synthase whose expression was confirmed by synthesis of HA in a microorganism that did not synthesize such a polysaccharide before.

1.5 Current Medical Study and Use

At the present time hyaluronan is an object of study in biochemistry, molecular biophysics, bioorganic and radiochemistry and the chemistry of polymer compounds. Medical studies of HA include its role in fertilization, embryogenesis, development of the immune response, the healing of wounds, oncological and infectious diseases, processes of ageing and the problems of aesthetic medicine.

The wide range of practical medical applications of hyaluronan continues to be based upon its anti-inflammatory, disinfectant and wound healing effects. HA promotes epithelial regeneration; prevents the formation of granulation tissue, adhesions and scars; reduces swelling and itching; normalizes blood circulation; promotes scarring of venous ulcers and protects internal eye tissue [24].

Hyaluronan is widely used in applied biochemistry and enzymology as a substrate for the quantitative determination of the enzyme hyaluronidase. Scientific disputes about the possible relationship between HA and hyaluronate lyase and the pathogenicity of some streptococci are to be, as it seems, permanently carried out. Currently, much attention is paid to the study of the secondary and tertiary structures and dynamic conformation of HA in aqueous solutions and biological fluids; the HA interaction with proteins, particularly receptor CD44 and other hyaladherins; and the HA biocatalytic cleavage with different hyaluronidase; the progression toward creating of recombinant strains and chimera products with the desired properties.

The theoretical research on polysaccharide biosynthesis with both bacterial and mammalian enzymes initially focused on gaining an understanding of the reaction mechanisms, but recently attention has been shifted toward solving the application problems, including the creation of recombinant strains and chimera organisms that could produce HA with the initially required properties.

In the last few years a very promising area of medicine and pharmacology has gained significant attention: the targeted delivery of drugs to specific organs or tissues. The development of nanotechnology has brought these studies to a new level because 'nanocontainers' have great potential to target delivery biologically active compounds to specific cells in the body. Liposomes, polymeric and lipid particles with a diameter of 100–300 nm, or nanocapsules of the same size, are typically used as nanocontainers. Hyaluronan could be used as an alternative approach in the creation of a macromolecular container. HA macromolecules could bind to the receptors on cytoplasmic membranes of various cells to target deliver biologically active compounds attached to the biopolymer carrier.

Using the technology of solid-state modification of polysaccharides that included the mutual action of super-high pressure and shear deformation, a group of Russian scientists have produced a number of unique products that contain hyaluronic acid. The new products could be considered drug-free (or non-drug) macromolecular therapeutics. These areas of research have made possible a broad range of new products based on the modified hyaluronan and could be used in the various fields of medicine.

1.6   Impact and Future Directions

Seventy-five years have passed since hyaluronan was discovered. In the world of scientific research, at this point past a substance's discovery, research is usually limited to a fairly narrow and specialized group of academics. However, the interest in this compound is far from decreasing; it is actually growing. Publications dedicated to the structure, synthesis, degradation, the biological role of HA and its application in various fields of chemistry, medicine and biology are on the constant rise. Between 1966 and 1975, 790 articles had been published; between 1976 and 1985 the number reached 2200, between 1986 and 1996 – 3300, between 1996 and 2006 – more than 7000. Two fundamental monographs [25,26] and a review [27] have also been published recently.

It is important to mention the most recent fundamental monograph – the series of five volumes under the title *Hyaluronan: From Basic Science to Clinical Application* by Balazs et al. [28]. The monograph covers many aspects of hyaluronan science and application, but what makes this book special is the comprehensive early history overview, followed by the chapters that describe contributions of the most outstanding scientists to hyaluronan chemistry and biology.

HA patents are also on the rise: between 1979 and 1987 USA patent bureau issued five patents regarding HA per year. In 1988 the amount reached a number 35. The same level was till 1995. In 1996 the sharp increase of the patents related to HA was observed, and the amount of the new patents is still increasing every year. As of February 2013, the number of patents involving the keywords 'hyaluronic acid' total almost 80 000.

In the beginning of the 1980s it was obvious that hyaluronic acid attracted attention of scientists in chemistry, biology, physics, medicine, and so on from all over the world. It became necessary to gather in order to discuss the latest results and perspectives. The first international meeting dedicated to HA took place in 1985 in Saint-Tropez (France). After that the conferences were held in London (UK) (1988), Stockholm (Sweden) (1996), Padua (Italy) (1999), Wales (UK) (2000), Cleveland (USA) (2003), Charleston (USA) (2007) and Kyoto, Japan (2010). In October 2004 The International Society for Hyaluronan Sciences (ISHAS, www.ishas.org) was created. Initially, it brought together scientists from 16 countries.

In June 2013 Oklahoma City (USA) was the host of the latest International Conference of Hyaluronic Acid. It brought together more than 200 scientists from 22 countries. Since then similar conferences have been carried out every two years, the next one will be in Florence (Italy) in 2015.

References

[1] Meyer, K., Palmer, J. (1934) The polysaccharide of the vitreous humor. *Journal of Biological Chemistry,* **107**, 629–634.
[2] Levene, P.A., Lopez-Suarez, J. (1918) Mucin and mucoids. *Journal of Biological Chemistry,* **36**, 105–126.
[3] Kendall F.E., Heidelberger M., Dawson M.H. (1937) A serologically inactive polysaccharide elaborated by mucoid strains of group A hemolytic *Streptococcus*. *Journal of Biological Chemistry,* **118**, 61–69.
[4] Duran-Reynalds, F. (1928) Exaltation de l'activité du virus vaccinal par les extrait de certains organs. *Comptes rendus des séances de la Société de biologie,* **99**, 6–7. Duran-Reynalds, F. (1929). The effect of extracts of certain organs from normal and immunized animals on the infecting power of vaccine virus. *Journal of Experimental Medicine,* **50**, 327–340.

[5] Balazs, E.A., Piller, L. (1943) The formation of the synovial fluid. *Magyar Orvosi Arch*, **44**, 1–11.

[6] Ragan, C., Meyer, K. (1949). The hyaluronic acid of synovial fluid in rheumatoid arthritis. *Journal of Clinical Investigations*, **28**, 56–59.

[7] Dorfman, A. (1948) The kinetics of the enzymatic hydrolysis of hyaluronic acid. *Journal of Biological Chemistry*, **172** (2), 377–87.

[8] Ogston, A.G., Stanier, J.I. (1951) The dimensions of the particle of hyaluronic acid complex in sinovial fluid. *Biochemical Journal*, **49**, 585–599.

[9] Laurent, T.C., Gergely, J. (1955). Light scattering studies on hyaluronic acid *Journal of Biological Chemistry*, **212**, 325–333.

[10] Fuoss, R.M. (1948) Viscosity function for polyelectrolytes. *Journal of Polymer Science*, **3**, 603–604.

[11] Balazs, E.A., Laurent, T.C. (1951) The viscosity function of hyaluronic acid as a polyelectrolyte, *Journal of Polymer Science*, **6** (5), 665–667.

[12] Caputo, A. (1957) Depolymerization of hyaluronic acid by X-rays. *Nature*, **179**, 1133–1334.

[13] Jiang, D., Liang, J., Noble, P.W. (2011) Hyaluronan as an immune regulator in human diseases. *Physiological Reviews*, **91**, 221–264.

[14] Linker, A., Meyer, K. (1954) Production of unsaturated uronides by bacterial hyaluronidases. *Nature*, **174**, 1192–1194.

[15] Balazs, E.A., Miller, D., Stegmann, R. (1979) *Viscosurgery and Use the Use of Na-Hyaluronate in Intraocular Lens Implantation.* Presented at the International Congress and Film Festival on Intraocular Implantation, Cannes, France, 1979.

[16] Higashide, T., Siguyama, K. (2008) Use of viscoelastic substance in ophthalmic surgery – focus on sodium hyaluronate. *Journal of Clinical Ophtalmology*, **2**(1), 21–30.

[17] Fraser J.R., Laurent T.C., Pertoft H., Baxter E. (1981) Plasma clearance, tissue distribution and metabolism of hyaluronic acid injected intravenously in the rabbit. *Biochemical Journal*, **200**, 415–424.

[18] Roseman S., Moses F.E., Ludowieg J., Dorfman A. (1953) The biosynthesis of hyaluronic acid by group A *Streptococcus*. Utilization of l-C14-glucose. *Journal of Biological Chemistry*, **203**, 213–225.

[19] Warren G.H., Gray J. (1959) Isolation and purification of streptococcal hyaluronic acid. *Proceedings of the Society of Experimental Biology and Medicine.* **102**, 125–127.

[20] Markovitz A., Cifonelli J.A., Dorfman A. (1959) The biosynthesis of HA by group A *Streptococcus*. VI. Biosynthesis from uridine nucleotides in cell-free extracts. *Journal of Biological Chemistry*, **234**, (9) 2343–2350.

[21] Lansing M., Lellig S., Mausolf A. *et al.* (1993) Hyaluronate synthase: cloning and sequencing of the gene from Streptococcus sp. *Biochemical Journal.* **289**, 179–184.

[22] Van de Rijn I., Drake R.R. (1992) Analysis of the streptococcal hyaluronic acid synthase complex using the photoaffinity probe 5-azido-UDP-glucuronic acid. *Journal of Biological Chemistry*, **267**, 24302–24306.

[23] DeAngelis P.L., Papaconstantinou J., Weigel P.H. (1993) Isolation of a Streptococcus pyogenes gene locus that directs hyaluronan biosynthesis in acapsular mutants and in heterologous bacteria. *Journal of Biological Chemistry*, **268**, 14568–14571.

[24] Radaeva I.F., Kostina G.A., Zmievski A.V. (1997) Hyaluronic acid: biological role, structure, synthesis, isolation, purification, and application (in Russian). *Applied Biochemistry and Microbiology* (Prikladnaya biokhimiya i mikrobiologiya), **33**, 133–137. Радаева И.Ф., Костина Г.А., Змиевский А.В. (1997) Гиалуроновая кислота: биологическая роль, строение, синтез, выделение, очистка и применение. *Прикладная биохимия и микробиология.* 33 (2), 133–137.

[25] Garg H.G., Hales C.A. (2004) *Chemistry and Biology of Hyaluronan.* Elsevier, Amsterdam.

[26] Kuo J.W. (2005) *Practical Aspects of Hyaluronan Based Medical Products.* CRC Press, New York.

[27] J. Necas, J., Bartosikova, L., Braune, P., Kolar J. (2008) Hyaluronic acid (hyaluronan): a review. *Veterinarni Medicina*, **53** (8), 397–411.

[28] Balazs, E.A., Hargittai, I., Hargittai, M. (2011) Hyaluronan: From basic Science to Clinical Application. *PubMatrix Inc.* Edgewater, New Jersey.

2

The Biological Role of Hyaluronic Acid

Many characteristics of hyaluronan are wonderful inventions of evolution. Such an evolutional definition could obviously be attributed to other biopolymers such as proteins, nucleic acids and polysaccharides. Indeed, evolutional potential, mostly related to biopolymers, strongly depends on molecular physico-chemical nature [1]. HA is considered one of the earliest evolutional forms of the polysaccharide family. So, let's look at hyaluronan's biological role in comparison with the molecular evolutions of other biopolymers.

2.1 Hyaluronic Acid Phylogenesis

From an evolutionary point of view, HA is considered quite a conservative polysaccharide since its chemical structure is completely identical for all known species at the different levels of the evolutional stairs. Such chemically evolutional conservatism indicates the importance of its biological functions. A comparison between the structures of the different polysaccharides 'located' at the different branches of the 'evolutional tree' leads to several interesting conclusions [2]:

- Alpha-D-glucans (glycogen, amylose and amylopectin) are usually intercellular back-up sources of energy and cells' building blocks. At the same time, beta-D-glucans are related to structural intracellular polysaccharides.
- Homoglucans and heteroglucans of the different families of the living nature are characterized by the monomers' stereoisomerism and characteristics of the nature of the bonds between them.
- Considerable amounts of hyaluronan, chondroitin sulfate and other glycosaminoglycans (Figure 2.1.) are found only in Chordata animals.

Hyaluronic Acid: Preparation, Properties, Application in Biology and Medicine, First Edition.
Mikhail A. Selyanin, Petr Ya. Boykov and Vladimir N. Khabarov.
© 2015 John Wiley & Sons, Ltd. Published 2015 by John Wiley & Sons, Ltd.

Glucosaminoglycan	Mol. weight	Disaccharide unit -[A-B]-	Comments
Hyaluronic acid	$4\,000 - 8 \times 10^6$	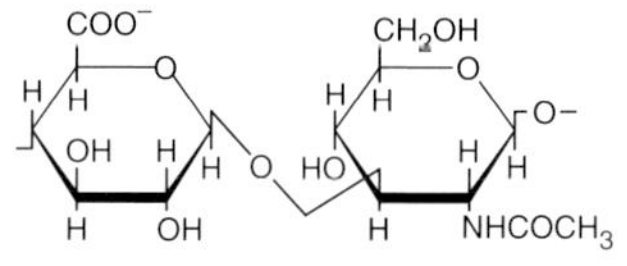 Glucuronic acid N-acetylglucosamine	Non-sulfated, non-protein-bounded polysaccharide found in synovial and eyeball liquid, vitreous body, cartilage
Chondroitin sulfate	$5\,000 - 50\,000$	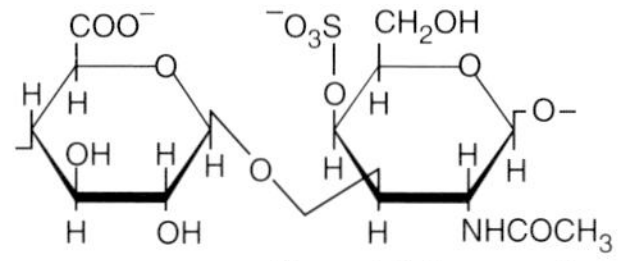 Glucuronic acid N-acetylglucosamine -4-(or 6)-sulfate	Sulfated, protein-bounded polysaccharide found in cartilage, cornea, bones, skin, arteries
Dermatan sulfate	$15\,000 - 40\,000$	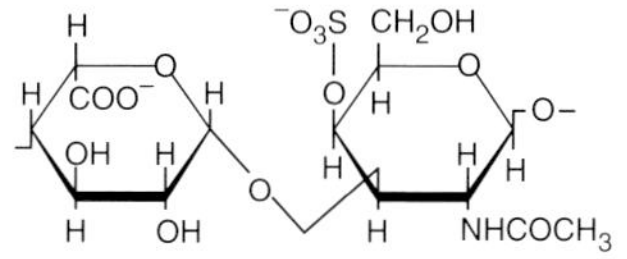 Iduronic acid N-acetylglucosamin -4-sulfate	Sulfated, protein-bounded polysaccharide found in skin, blood vessels, heart, heart valve
Heparan sulfate	$5\,000 - 12\,000$	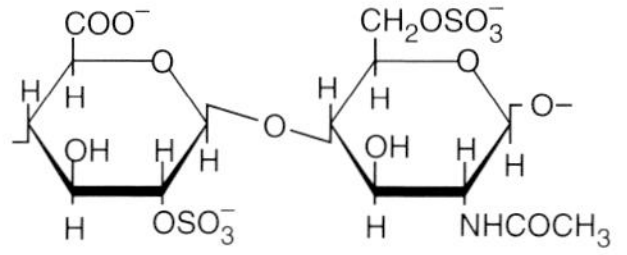 Glucironate -2-sulfate N-acetylglucosamine -4-sulfate	Sulfated, protein-bounded polysaccharide found in lungs, arteries, bazal lamina, cell membranes
Heparin	$6\,000 - 25\,000$	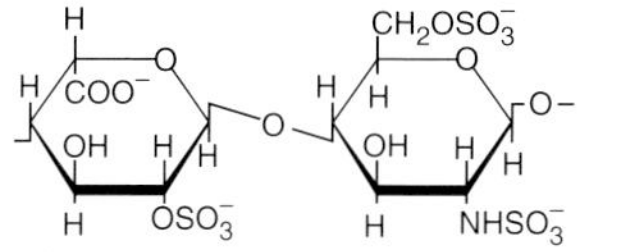 Iduronate -2-sulfate N-sulpho-glucosamine -6-sulfate	Sulfated, protein-bounded polysaccharide found in lungs, liver, skin, corpulent cells
Keratan sulfate	$4\,000 - 19\,000$	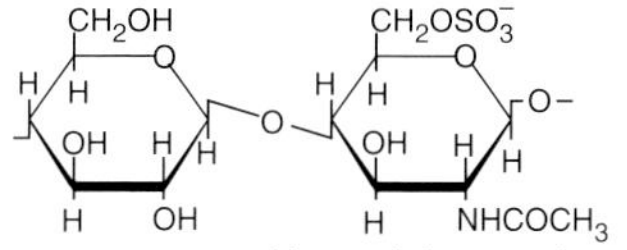 Galactose N-acetylglucosamine -6-sulfate	Sulfated, protein-bounded polysaccharide found in cartilage, cornea, intervertebral disk

Figure 2.1   Structure of major glycosaminoglycans

In the modern system of the organic world, main groups of living organisms are differentiated by the nutrition mode, which is mainly determined by polysaccharides and is located on the external cell walls. Thus, the cellulose cell wall allows the living system to consume only water, inorganic compounds and gasses. Such cells acquired the ability for photosynthesis of the organic compounds, that is, autonomous nutrition type, which is a key factor in the evolution of the plant kingdom. Other cells in the process of evolution acquired the cell wall made from the polysaccharide chitin, which provides an ability to consume the organic compounds of high molecular weight. Such organisms are able to use the nutrition of decomposed organic materials (saprophytic nutrition) and give origination to the kingdom of fungus [3]. The chitin is widely seen in the branch of the invertebrate animals, while HA and related glucosaminoglycans are found in the branch of Chordata animals. However, the evolutional role of HA and other glycosaminoglycans is still not quite clear.

Apparently many features of the biological systems could be understandable based on the properties of the macromolecular polysaccharide components and their relative biochemistry. Based on the known physico-chemical properties of HA we can try to understand its role in the biological evolution of the species. Or, in other words, we will try to determine when and how HA arrived on the 'life tree' and what new things it has introduced to the process of evolution according to its physico-chemical and physiological properties.

In 1924, based on the compound's properties, processes and rules in organic, physical and biological chemistry known at that time, A.I. Oparin formulated an understanding about self-organization of the molecular structures that could have led to the arrival of the life on Earth [4–11]. The problem of origination of the macromolecules and their molecular evolution was subject to active development impulse after 1953, when S. Miller, H. Uri and later, other scientists (T.E. Pavlovskaya and A.G. Pasynsky, 1957), experimentally confirmed the possibility of biologic monomer synthesis (amino acids, organic acids and other organic compounds) from the gases of the supposed primary atmosphere of the Earth (H_2, CH_4, NH_3 and H_2O vapour) under electric charges. It was found that polymerization of the amino acids and formation of the proteins could occur quite easily under unbiogenic conditions at high temperature [12,13]. Due to easy polymerization of amino acids, molecular evolution at the early stages could develop in the absence of nucleic acids that require a more complicated system for precursor biosynthesis and polymerization. This is why the 'protein first' hypothesis [14] was formulated as opposed to the 'gene first' hypothesis [15]. To dismiss a 'chicken and egg' discussion, we would like to point out an evident fact. In the process of molecular evolution, it was found that the non-biogenic matrix less synthesis of the proteins was replaced by biogenic matrix synthesis under control of the nucleic acids codons (genes). At this point the 'genealogic tree' of the protein molecule starts to form [16].

A study of the amino acid sequence of the same proteins isolated from the animals, which occupy difference levels of the evolutional stairs, allowed for the rate of evolution of the proteins and their genes to be determined. The evolution of some proteins, such as globine (haemoglobine), cytochrome C and fibrinopeptide is one of the best examples for this task [17,18]. Evolution of the proteins in these life systems is carried out step by step and is related to changes of purine and pyrimidine bases in the nucleic acids, that is, the mutations in DNA. Contrary to non-biogenic synthesis, the biogenic matrix synthesis of the proteins became stereospecific, which means the proteins are synthesized only from amino acids of L-forms. Due to the L-configuration of the amino acids, the side chains of the protein backbone create an environment in which steric hindrance is minimal [18]. This protein polymer's property is important because the biological activity and functions of proteins are mainly related by the three-dimensional configuration of the polymer chain. This configuration is a result of the amino acid sequence and orientation in the primary, secondary and tertiary structures of proteins. All proteins and nucleic acids are linear polymers and include the standard set of monomers – 20 amino acids and five nucleotides connected with the standard peptide and phosphoroester bonds correspondingly. Polysaccharides on the other hand, can be formed from only one carbohydrate residue (homopolysaccharides) or from several different residues (heteropolysaccharides).

Both proteins and nucleic acids are representative of the 'biological unity' of a living system. The polysaccharides largely reflect its 'diversity' [18]. Thus, the extracellular polysaccharides cellulose and pectin are well represented in the kingdom of plants; chitin in the branch of invertebrate animals and hyaluronan in the branch of Chordata species. Synthesis of polysaccharides in non-biogenic conditions is more difficult than synthesis of the proteins and nucleic acids, which is why appearance of the gene of the hyaluronan synthase, followed by appearance of hyaluronan, is highly possible. This sequence of events means that this polysaccharide originated in evolution by biogenic route as a result of the formation of the corresponding gene (genes) of hyaluronan synthase – the enzyme responsible for synthesis of hyaluronan. Indirect evidence for such a hypothesis could be provided by experiments that model the non-biogenic reactions of the polymerization of glucose. Thus, the gamma-irradiation of the aqueous solutions of glucose resulted in polymers that contain many non-typical crosslinks, but the initial carbohydrate monomer is decomposed due to action of OH-radicals and solvated electrons that originated as a result of the radiolysis of the glucose aqueous solution (see Section 4.3). As a result, the polysaccharides contain the carbonyl groups and gamma-lactones. This polymeric product characteristic is maintained even after the acid hydrolysis and oxidation with periodate [19]. Similar polymerization of polysaccharide in the aqueous solutions could be carried out under UV irradiation [20]. However, in the polysaccharides of the biological origin, the monomers are usually not connected between them with ester bonds or C-C bridges, but rather have glycosidic bonds. Similar to other model reactions of non-biogenic condensations (amino acids and nucleotides), the polymers, which are based on the monosaccharaides, are better formed in the thermal reactions. In hexose (particularly in glucose) the primary hydroxyl at C_6 has higher reactivity than the secondary hydroxyl groups at C_2, C_3 and C_5. Because the reactivity of glycosidic hydroxyl at C_1 exceeds all other hydroxyls, the activation energy of the second, third and fifth hydroxyls cannot be reached at the reaction temperatures 141–143°C. So, in the formed polymer, the 1,6-glycosidic bonds are predominant [21]. At higher temperatures, a considerable amount of insoluble gels is formed and a significant amount of ether crosslinking occurs. The hydrolysis of such polysaccharides leads to formation of glucose, as evidence of the absence of C-C type cross-linking. At approximately 155°C, the cross-links of the ether nature are not formed [21]. The result of the periodate oxidation of the polysaccharide polymers is in agreement with the general rule whereby the type and frequency of the specific type of bond in the polymer is determined by the reactivity of the specific hydroxyl at the specific reaction temperature [18]. Based on that rule, one could assume that the evolutional precursors of glucagon and starch (alpha-D-glucans) originated by non-biogenic route as a result of the thermal reactions and were included into the primary biologic objects. The secondary hydroxyl at C_3 has the lowest reactivity (the highest activation energy), and so, consequently, the synthesis of hyaluronan with beta-1-3-bonds, even in non-biogenic conditions, is difficult. That is why it is possible to assume that HA arrived later than other biopolymers on the 'molecular evolution stairs': likely when the gene capable of creating a beta-1-3-bond between D-glucuronic acid and N-acetyl-D-glucosamine appears.

2.1.1   Polysaccharide Structure and the Problems of Phylogenesis

In order to understand the place and time of the appearance of hyaluronan in phylogenesis (the historical development of life), we must pin-point its main milestones. It is assumed that the first living cells appeared on Earth approximately 3.5 billion years ago [22]. An era where 'life' and 'proto-life' co-existed came to a close approximately 1.8 billion years ago [23]. The organisms capable of photosynthesis appeared and the content of biogenic oxygen in the atmosphere reached the 'Pasteur point' (0.01 form the current level), meaning that the 'primary bullion' had been exhausted forever. The non-biogenically synthesized organic compounds were spent due to heterotrophic organisms and oxidation by the atmospheric oxygen. At the same time, the rules of the symbiotic union of the genetic material of the symbionts came into action, increasing reliability and survival of the new cells that united the partners in the symbiosis. As a result, approximately 1.4 billion years ago, eukaryotic cells appeared. They are found in the fossil traces from the late Precambrian rocks [24].

Obviously, large gaps in our understanding of phylogenesis can be filled only by unsupported reasoning. Nevertheless, fossil traces and comparative biochemical studies of molecules and organisms at different levels of 'evolution stairs' provide sufficient information to support a hypothesis about main stages and mechanisms of evolutional development of living systems. In condensed form, these stages are presented in a bio-geochronological table (Table 2.1). Next,

Table 2.1   *Bio-geochronology and evolution of chordate animals*

Era	Period	Beginning, millions of years ago	Duration, millions of years	Time of chordate animals appearance
Cenozoic	Quaternary Neogene	70		Evolution of human being Divergence of mammals
Mesozoic	Cretaceous	135	65	Placental mammals
	Jurassic	180	45	Birds
	Triassic	225	45	Non-placental mammals
Paleozoic	Permian	270	45	
	Carboniferous	350	80	Reptiles
	Devonian	400	50	Amphibians
	Silurian	440	40	Bony fishes Cartilaginous fishes
	Ordovician	500	60	Cyclostomatousoka chordates
	Cambrian	600	100	Jawless chordates.
Proterozoic				Appearance of eukaryotic cell approximately 1.4 billion years ago
Archean				Appearance of first prokaryotic cells approximately 3.5 billion years ago

we need to determine the approximate time of appearance the hyaluronan synthase gene, which is the enzyme responsible for synthesis of HA.

The evolutional development and appearance of a variety of species took place as a result of the genetic rules of evolution. The evolution of genomes went in a direction wherein the amount of the genetic information was increased, that is, an increase of DNA content. Thus, mycloplasm, the simplest bacterial cell without a polysaccharide external layer, has only 0.5×10^6 pairs of nucleotides in its DNA and can code about 470 general proteins. The bacteria with polysaccharide cell shells (murein) have 3×10^6 pairs of the nucleic bases in its DNA. Monocellular eukaryotic cells, like yeast, include about 15×10^6 pairs of the nucleic basis.

A multicellular organism requires up to 80×10^6 pairs of the nucleic bases in order to exist. In fish cells, genetic material contains about 300×10^6 pairs of nucleic bases whereas in a mammalian cell there are about 3000×10^6 pairs. The human genome contains 3164×10^6 pairs of the nucleotides with the number of genes from 20 000 to 27 000 (according to the results of the 'Human Genome' program). Thus, during evolution the amount of genetic material and new genes responsible for synthesis of new biopolymers steadily increased. In the process of molecular evolution, newly arrived biopolymer structures 'look' for their functions. As is known, the natural selection occurs at the functioning level [25]. Different extracellular polysaccharides probably played an important role in the species' divergence.

A 1969 review on the geochemistry of fossil hydrocarbons contains information that determines chitin was found in the residues of marine invertebrates [26]. One of the products of HA breakdown, glucosamine, was found in the bonds of fossil fishes [27]. The large amount of fossil biochemical materials could be found in the Cambrian sediments (about 500–570 million years ago) as well as in more recent layers [12,18,28]. The first cartilaginous creatures such as cyclostomata and cranial chordates, which contained considerable amounts of hyaluronan in their tissues, appeared about 500 million years ago in the early Ordovician period [16,18,29,30] (Table 2.1) and blossomed during the Silurian and Ordovician periods. In that period of time an evolutional explosion in the number of marine animals is observed. Evolutionally older animals such as coelenterates, worms, molluscs and arthropods, do not contain hyaluronan in considerable amounts. Thus, there is a quite clear boundary that makes it possible to supposedly determine the time of the appearance of the gene of hyaluronan synthase. It preceded or coincided with the appearance of primitive cartilaginous creatures, which could be dated back to the beginning of the Cambrian period of Paleozoic era. Hence, it can be assumed that the gene(s) for hyaluronan synthase and HA itself appeared in either the pre-Cambrian or early Cambrian period about 570 million years ago. Since that time, the rate of evolution increased significantly, resulting in evolutional explosion in the branch of cartilaginous animals known as the 'Cambrian Explosion'. Another stream of indications that the gene of hyaluronan synthase first appeared during the period of the phylogenesis lies in the law of biogenesis.

By comparing embryos of different classes and groups of chordates, K. von Baer had postulated a law of embryonic similarity, which states that embryos of the same phylum of animals are very similar at the initial stages of the developments. Later (in 1866), on the basis of these and other data, E. Haeckel formulated a basic law of biogenesis that states ontogenesis is a short repetition of phylogenesis. Early stages of animal and human embryonic development preserve some features characteristic of primitive chordate species. This happens because one of the main integration mechanisms of early stages of development are the mechanisms of embryotic induction according to which embryonic structures, such as cord, nerve tube and somites, represent organization centres that determine further ontogenesis. Thus the cord, for instance, induces development of embryo's nerve tube and spine [3]. According to current understanding, the basis of such interrelationships is a genetic control of development by the general genes of ontogenesis regulation inherited from common ancestors. The gene that encodes hyaluronan synthase and is responsible for synthesis of HA is probably one of such genes, since HA is known to participate in the formation of the main proteoglycan of the cord. Synthesis of HA itself requires availability of nucleotide derivatives of sugar precursors, such as uredines phosphohexoses (UDP-glucuronic acid and UDP-N-acetyl glucosamine), which could be seen in glucose metabolic pathways (Figure 2.2).

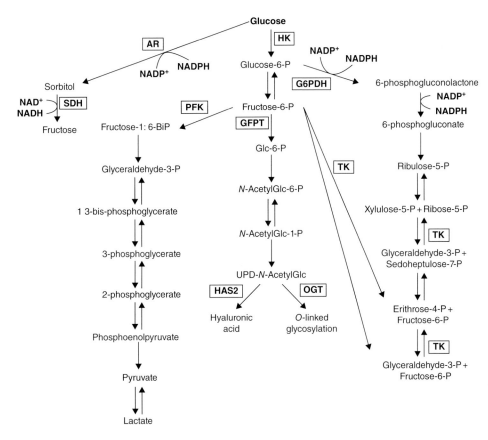

Figure 2.2 Biosynthesis of HA, its sugar precursors and other glycosaminoglycans from glucose [31]. Rate-limiting or important enzymes: AR – aldose reductase; ECM – extracellular matrix; G6PDH – glucose-6-phosphate dehydrogenase; GFPT – glucosamine:fructose acetyl transferase; HAS2 – hyaluronan synthase-2; HK – hexokinase; OGT – O-linked glycosylation transferase; PFK – phosphofructokinase; SDH – sorbitol dehydrogenase

It is also possible that the gene encoding hyaluronan synthase had been formed much earlier but its expression was inhibited or completely blocked. One can assume that active *in vivo* synthesis of HA and related glycosaminoglycans contributed significantly to new directions and acceleration of evolution processes. The evolution was further directed to lead to an increase in size and shapes of chordate species and also their transition onto the land. Without HA, which retains large amounts of water in skin, moving onto the land would be problematic. The branch of invertebrates having chitin skeletons includes a great variety of species (about 2 million arthropods) but is limited in cell reproduction and organism size. At the same time, the branch of chordates with cartilaginous or bone skeleton and connective tissues that contain a significant amount of HA, were successful in increasing multicellularity, the organism's size, organization complexity, as well as interaction between cells, tissues, organs and organ systems. It thus appears that the variety of a vertebrate's cells and tissues rich in HA and related glycosaminoglycans contributed to the acceleration of the evolutionary process and the complexity of reproduction and growth. Invertebrate animals are in part characterized by limitations in growth and number of cells (worms, arthropods). The limitations are directly associated with chitin-based exoskeletons. In the branch of chordate animals, chitin is not present and intercellular relationships are to a large extent mediated by hyaluronan and other related polysaccharides.

However, in the light of the aforementioned, rare exceptions of HA in bacteria, protozoa, molluscs and plants, as reported in the literature, require certain explanation. The ability of more ancient animals and plants to synthesize HA is due to the presence of the gene encoding hyaluronan synthase. In examination of this connection, one ought to review genetic mechanisms of new genes and genome formation.

Currently, a range of mechanisms of the eukaryotic genome's structural evolution is known. Polyploidy is in this range. Its contribution is probably unimportant however, since by this mechanism one would expect to see a jump-like revolutionary increase in the volume of genetic material. In reality, there was a smooth evolutionary accumulation of DNA material from less organized to highly organized species on the evolution tree. It is thought that an increase in the DNA amount in eukaryotic cells occurred through amplification, for example, formation of certain copies of DNA and divergence in nucleotide sequences due to accumulation of mutations in new genes. According to estimates obtained in the evolution studies, useful mutation in one medium-size gene occurs once every 200 000 years. One of the noticeable examples is the protein haemoglobin. It has been shown that about 500 million years ago, during evolution of superior fish, certain mutations resulted in duplication of the haemoglobin gene [16,17]. Similar phenomena are known to be responsible for appearance of gene of collagen [17], a protein functioning in extracellular matrix in association with hyaluronan. Possibly, the genes encoding hyaluronan synthase evolved the same way. The report [32] provides comparative analysis of the sequences of nucleotides in the family of genes and amino acid sequences in the hyaluronan synthases HAS_1, HAS_2 and HAS_3 from amphibians, birds and mammals (e.g. species whose appearance dates back to different period of history). The authors came to the conclusion that the synthesis of HA in vertebrates is regulated by a relatively old group of genes that appeared through duplication and divergence of an even older parent gene [32]. Three hyaluronan synthase genes are localized in different chromosomes, thus supporting the theory that they appear through duplication of the ancient gene.

Recombination events play an important role in the formation and evolution of new genes. Other mechanisms such as duplication, transposition, reversed transcription, incorporation of labile genetic materials in the animal DNA and tandem duplication, all behave in the same evolutionary direction [33]. As a result of the differing mechanisms, a new complex gene becomes available from simple genes. One example of gene elongation through tandem duplication is the gene responsible for collagen synthesis [34]. Such processes in genome evolution have led to the gene's divergence, multiplication of gene families, and variety of the protein structures they regulate. Changes in the structure of genomes are transferred from one generation to another or in other words, vertically. However, today we know that the lateral transfer of genes exists between organisms through the mechanisms of conjugation, transformation, and transduction. Leading roles in the lateral transfer of genetic information belongs to viruses and phages. They can transfer foreign DNA in various eukaryotic cells where it can be stored as an extrachromosal element or integral expressible part of a gene. The possibility of lateral transfer of genetic material can serve as explanation for the presence of gene encoding synthesis of HA in the animals occupying lower levels on the evolution stairs compared to chordate species that are assumed to form the first gene of hyaluronan synthase. The single cases of existence of HA are described in the literature for invertebrates such as mollusc *mytilus galloprovincialis* [35] and chlorella-related single-cell green algae [36]. For the latter species, the lateral transfer of hyaluronan synthase gene into eukaryotic cell of green algae was experimentally confirmed using a virus. After 30 min of infecting the cell with the virus PBCV-1, the algae starts synthesizing HA identical to that in vertebrae animals. The gene of viral hyaluronan synthase shares 28–30% of homology with the hyaluronan synthase gene of vertebrates [37]. This is a clear example of lateral transfer of genes between eukaryotic cells. It is much harder to explain the presence of the hyaluronan synthase gene in prokaryotic cells that produce extracellular HA homologous to that in vertebrate animals. Such ability is known for gram-positive *Streptococci* bacteria (groups A and C) [38] and gram-negative *Pasteurella multocida* bacteria [39].

These examples ought to be reviewed through the prism of parasitism. The microorganisms capable of synthesizing HA in general are pathogens of animals and humans. It becomes obvious that by acquiring the hyaluronan synthase gene, these bacteria obtained an ability to cross cellular and humoral immunity of the host organism. Indeed, once the hyaluronan capsule is removed from the cells of *Streptococci* bacteria by the hyaluronidase enzyme, the pathogen potential of the bacteria cells is reduced 10 000-fold [40], suggesting effective recognition and elimination of the pathogen by immune system of the host. If bacteria dressed in the HA capsule present themselves as wolves in lamb skin, then many eukaryotic parasites of vertebrate animals can cross the immune barriers by destroying the hyaluronan shield with the use of the hyaluronidase enzyme. This is how a single-cell eukaryotic parasite *Balantidium coli* acts in a human organism. The bacteria has probably acquired the hyaluronidase gene and used the enzyme to break down the intestinal mucosa barrier in the human body. The third strategy, used by parasites to adapt to the host with the help of polysaccharides, is to build up chitin shells. The vertebrate animals do not have enzymes or immune systems capable of chitin recognition and breakdown, which is why the world of 'chitin parasites' composed of fungi, helminths and arthropods, is so abundant. The phenomenon of parasitism is as old as life itself. Noticeably, the pathways of parasite adaptation include differences in the extracellular polysaccharides.

2.1.2   Physico-Chemical and Functional Differences of Polysaccharides

As was mentioned, beta-D-glycans, to which hyaluronan is related, are extracellular or intercellular polysaccharides. It is important to find out what new HA and related glycosaminoglycans have been introduced into the intercellular relations of chordate animals. To do that, let's look at the differences of the chemical structures and physico-chemical properties of polysaccharides with the purpose of understanding what physiological functions they perform.

Functionally, the polysaccharides are divided into three main groups. The first group is intercellular polysaccharides, which fulfil energetic functions (starch in the plant world and glycogen in the animal world). The reserve functions of these polysaccharides are dictated by their polymer nature. They are less soluble than monosaccharides and therefore do not affect osmotic pressure in the cell. Because of that, they can be accumulated in the cells and if necessary can serve as a source of glucose for energy and building functions. Chemically they are mainly homo-polysaccharides, meaning all monomers are identical and connected with alpha-1-4-glycosidic bonds with some presence of alpha-1-6-glycosidic bonds in the branching points. There are hydrolytic enzymes in the cell capable of cleaving these bonds and releasing monomeric sugars.

The second group is structural polysaccharides, the main function of which is to provide the cells, tissues and organs with mechanical rigidity. The typical compound related to this group is the extracellular plant polysaccharide cellulose. This is a linear homo-polysaccharide built from the glucose residues. Because of such structure, extracellular cellulose is similar to the intracellular polysaccharides starch and glycogen. However, there is a very important difference between these homo-polysaccharides: In glycogen and starch the 1-4-glycoside bonds are in alpha configuration, but in cellulose they are in beta configuration. Such a difference in structure leads to essential differences in their secondary structure and functional properties. Due to geometrical features of alpha-1-4-bonds, the linear fragments of the glycogen and starch chain adopt a banded helical conformation. As a result, dense and compact granules occupying minimal volume in the cell are formed.

The beta-configuration of the 1-4-bonds in the cellulose molecule leads to extended polymer chains that form long and water-insoluble fibrils as a result of intermolecular hydrogen bonding. Further, while the alpha-1-4-bonds of glycogen and starch are hydrolysed with animal enzymes (alpha-amylases), the beta-1-4-bonds in the molecule of cellulose are not hydrolysed, since animals do not have the corresponding gene for cellulase, the enzyme that hydrolyses cellulose.

Another linear polysaccharide of the group of the structural polymers is chitin. It is built from the residues of N-acetyl-D-glucosamine, connected also with the beta-1-4-glycosidic bonds. In the aminosugar molecule of N-acetyl-D-glucosamine, the second carbon atom is attached to acetylated amino groups instead of hydroxyl groups. In the biopolymers, the amino group of the aminosugar is usually acetylated, resulting in eliminations of their positive charge.

The long parallel chains of chitin are arranged in tufts similar to those of the cellulose chains. Chitin is involved in the composition of cuticles, the external skeletons of invertebrate animals and cell envelops of fungus, where it is usually bound to proteins, lipids, pigments, inorganic salts and most typically, potassium carbonate [2].

The third group are polysaccharides involved in the intercellular matrix. Contrary to cellulose and chitin, the polysaccharides of the extracellular matrix are water-soluble and strongly hydrated. Hyaluronan and other glycosaminoglycans belong to this group (Table 2.1). Compared to other extracellular polysaccharides (cellulose, chitin) they possess significantly wider functions and actively participate in cell proliferation, migration, differentiation, and recognition.

The main branches of the evolutionary 'species tree' clearly differ in their polysaccharide composition. The most ancient structural polysaccharides are most likely murein-forming cell walls of bacterial cells. The structural basis of murein comprises long parallel-oriented polysaccharide chains composed of N-acetyl-D-glucosamine and N-acetyl muramic acid linked by beta-1-4-bonds. The chains of polysaccharides are cross-linked by short polypeptide fragments of various compositions in different bacteria.

The polysaccharides of the aforementioned three groups are characteristic of eukaryotic cells. In the first group of reserve polysaccharides, which have alpha-1-4- and alpha-1-6-glycosidic bonds for energy storage, clear differentiation between different evolutionary branches of eukaryotic species is observed. The plant kingdom has chosen starch, while the animal kingdom prefers glycogen as its reserve polysaccharide. These two polysaccharides share a lot of structural similarity. For instance, glycogen is a structural analogue of starch amylopectin; the only difference is the higher degree of branching in glycogen. In glycogen, each 10-residue long fragment has a 1-6-glycosidic bond, while in starch amylopectin one 1-6-glycosidic bond is found for every 20–25 glucose residues. This seemingly nonessential chemical difference has a profound physiological implication, as glycogen forms more compact structures that occupy smaller volume in the cell. The size of glycogen granules is in the nanometre range (10–40 nm), whereas those of starch granules is in the micrometre range (15–50 μm). At the same time, the molecular weight of glycogen and starch are comparable, that is, from 300 000 to 100 000 000 for glycogen and 100 000–100 000 000 for starch [41]. A higher degree of glycogen branching creates another functional advantage for animal cells: highly branched structures present higher numbers of terminal monomers which improves enzymatic turnaround either upon hydrolysis or glycogen synthesis, since multiple enzyme molecules can act on multiple branches at the same time.

Just like glycogen, starch amylopectin is composed of glucose residues linked together by alpha-1-4-glycoside bonds. At the branching points, glucose residues are linked by alpha-1-6-glycosidic bonds. This leads to the formation of the tree-like structure. Starch also comprises the linear polysaccharide amylose, which is composed of 200–300 glucose residues linked by alpha-1-4-glycoside bonds. This structure is thanks to the alpha-configuration of the glucose residues that form helix-like spatial configurations. Arrangement of these two polysaccharides in the form of starch granules leads to an increase in granule size and differentiation in shape. The shape of the glycogen can be described as spherical. Nanometre size, spherical shape of glycogen macromolecules and high metabolic rates (both synthesis and breakdown), are functional properties well suited to mobile animal cells.

By considering glycogen and starch as examples, one can clearly see how the chemical, physico-chemical, and conformational properties of biopolymers have been transformed into their functional characteristics according to the logic of cellular evolution. The second group of polysaccharides with beta-1-4-glycosidic bonds and well-defined structural functions is evolutionary diverse as well: the cellulose is found in plants, whereas chitin is typical for invertebrates and fungi. The third group of polysaccharides, glycosaminoglycans family comprising HA is characteristic to the chordate animals. Thus, even from such a brief review of polysaccharide distribution in living organisms, it is clear that polysaccharides have played an important role in the branching of the evolution tree and the emergence of a variety of species. It is quite obvious that HA participated in the evolution of the highly organized branch of chordate animals. Let's now examine what kind of functional novelty HA brought to the evolution process.

2.1.3   Biochemical Features of Hyaluronic Acid and Other Glycosaminoglycans

All glycosaminoglycanes have several common features.

- All glycosaminoglycans are linear non-branched polysaccharides.
- All glycosaminoglycans are built from repeated disaccharide units and thus are heteropolysaccharides (Figure 2.1) contrary to homopolysaccharides with identical monomers (starch, glycogen and cellulose are built from the glucose monomer and chitin is built from the acetyl amino glucose monomer).
- All glycosaminoglycans (with the exception of hyaluronan) contain sulfate groups as O-esters or N-sulfates.
- The amino group of the amino sugars is usually acetylated, which leads to the disappearance of the positive charge.
- With the exception of HA, all glycosaminoglycans are covalently bound with proteins and this leads to the formation of proteoglycan aggregates. Hyaluronan can be present not only bound to other components, but also as a free molecule.

Thus, even based on the chemical composition, the family of glycosaminoglycans differs from other polysaccharides. The important feature of glycosaminoglycans that differs from other polysaccharides is solubility in water. Other polysaccharides are partially soluble or insoluble in water. Hyaluronan could be additionally distinguished from the other members of the family by the following features: it is not sulfated, is not chemically modified after synthesis, it has high molecular weight and it can be found in a free state. Another interesting HA feature is its method of synthesis. Hyaluronan is synthesized by hyaluronan synthase, which is connected with the cytoplasmic cell membrane and is removed through the membrane directly on the outer cell surface into extracellular matrix during chain elongation. (It is interesting that the polysaccharide cellulose chains in the plant cells are synthesized in a similar way [42].)

It is believed that HA is the earliest evolutional form of glycosaminoglycans. Contrary to HA, other glycosaminoglycans are synthesized in the Golgi apparatus. They covalently bond with proteins to form the proteoglycans, which further transfer into the extracellular matrix by the exocytosis mechanism [41]. Three forms of hyaluronan synthase (HAS_1, HAS_2, HAS_3) are found in human and vertebrate bodies. The enzyme HAS_1 performs slow synthesis of high molecular weight HA, the enzyme

HAS_2 is more active and synthesizes polysaccharide with lower molecular weights up to $2\,000\,000$ Da. The enzyme HAS_3 is the most active but can only synthesize short HA chains from $200\,000$ to $300\,000$ Da [32,43,44].

The aforementioned three hyaluronan synthases are the products of three different genes. As was shown through experiments with mice, the most important from the evolutional point of view is the gene HAS_2. Mice with HAS_1- and HAS_3- deficiencies were viable but those with HAS_2 were not. Such mutation in gene HAS_2 is lethal during the embryonic development. The embryos were found to be almost completely lacking HA and had serious deficiencies incompatible with life [32]. In evolutionary terms, the role of three hyaluronan synthases with different activity kinetics and HA of different molecular weight remains unclear.

As was mentioned previously, hyaluronan is non-branched, non-sulfated glycosaminoglycan consisting of 5000–30000 disaccharide units formed by N-acetyl-D-glucosamine and D-glucuronic acid connected with beta-1-3 glycoside bond. Disaccharides are connected with beta-1-4-glycosidic bonds (Figure 2.1). Hyaluronan can reach the molecular weight of 8×10^6 Da, which significantly exceeds the length of other glycosaminoglycans, the molecular weight of which is within the range of 4000–50000 Da (Table 2.1). All glycosaminoglycans are water soluble. The polysaccharide chains of HA are sterically more rigid structures compared to, for example, protein's polypeptide chains. That is why the molecules of HA cannot form extra-compact globular structures characteristic of proteins.

In conditions with a pH around 7.0, the carboxylic groups of HA are dissociated and the polymer molecules have high-density negative charges. They attract sodium, potassium, magnesium, calcium and other osmotically active cations. Because of that, HA can bind up to 1000 times more water than the weight of the macromolecule itself. Due to these physico-chemical properties, the HA molecule adopts an enlarged conformation, occupies an extremely large volume relative to its mass and forms gels even at very low concentrations.

Depending on hyaluronan concentration and molecular weight, biopolymer gel media determines the physiological functions of the cells, tissues and organs, in functions of water-binding, ionic exchange, molecule's size-dependent diffusion and impermeability towards large molecules and cells. As a consequence, it provides protection from penetration of high molecular weight toxins and microbiological invasions. The presence of the large amount of HA-bound water creates a swelling pressure (turgor) that resists compression forces as opposed to collagen fibres that resist extension forces. The viscoelastic properties of hyaluronan in the synovial liquid of joints allow the biopolymer to act as a lubricant and shock absorber. After synthesis, HA is not subjected to additional chemical modification, does not contain chromophore groups and does not absorb in the visible range of the light spectrum. The aqueous solution of the polysaccharide is completely transparent and used for filling of the vitreous and other eye structures. Each of these physico-chemical characteristics and related biological functions of hyaluronan and other glycosaminoglycans have shaped the complex of new functional properties to assure accelerated evolution in the branch of chordate animals.

A brief review of molecular evolution of polysaccharides makes it possible to draw some general conclusions: (1) Extracellular polysaccharides such as cellulose, chitin and hyaluronan participated in the branching of the evolution tree of species, (2) cellulose participated in the formation of the plant world, (3) chitin was involved in the evolution of the branch of the invertebrate animals, (4) HA contributed to the evolution of the chordate animals, (5) all extracellular polysaccharides include beta-D-glucans and (6) extracellular polysaccharides containing beta-1-3 glycoside bonds likely appeared by a biogenic route after the genes encoding the enzyme capable of forming beta-1-3-bonds emerged as a result of genomic evolution of eukaryotic cells.

It is interesting that the enzymes responsible for the synthesis of the main extracellular polysaccharides (i.e. hyaluronan and cellulose) are localized in the cytoplasmic cell membrane. During the synthesis, a growing polymer chain diffuses through the membrane directly into extracellular space. Such a synthesis route is apparently more ancient and differs from the synthesis of other extracellular polysaccharides that are synthesized inside the cell and transported into extracellular media by the exocytosis mechanism.

2.2 Functions of Hyaluronan in Human Ontogenesis

The last few decades witnessed a growing number of publications on hyaluronan's role in fertilization (in direct correlation with the development of numerous programs of *in vitro* fertilization), cell division and migration, angiogenesis, wound healing and tissue regeneration. It is known today that HA is actively involved in the regulation of cell division, migration, differentiation and tissue and organ regeneration at all stages of organism development or ontogenesis. There are three periods in ontogenesis: (1) the period of embryonic development starting from fertilization up to the birth, (2) the period of growth and sexual maturity and (3) the ageing period. The function of HA is particular during all three periods. Most HA is synthesized in the embryonic period. In chordate animals the regulatory type of embryonic development dominates [3] and the entire process includes three stages: (1) cleavage of zygote and formation of the blastula, (2) gastrulation or formation of germ layers and (3) histo- and organogenesis or formation of tissues and organs. HA is actively involved in all stages of embryogenesis.

2.2.1 Role of Hyaluronic Acid in Fertilization

When an egg cell leaves the ovary, it is coated by two layers [45]. The external surface of the cytoplasmic membrane is covered with *zona pellucida*, a coat composed of a noncellular, jelly-like substance. The surface of the *zona pellucida* is populated by many layers of the follicular cells that compose the layer called *corona radiata* (Figure 2.3). Both *zona pellucida* and *corona radiata* contain significant quantities of hyaluronan in the extracellular matrix. It has been experimentally shown that HA with a molecular weight of 2–5 million Da is synthesized by hyaluronan synthase [46]. Electron microscopy revealed that long molecules of HA in the intercellular matrix of the *corona radiata* are entangled and organized in the form of fibres forming a homogeneous network anchored on the surface of the cells. Cells synthesize the protein CD44, the receptor for HA on the surface of follicular cells. The organization of hyaluronan in this structured matrix is accomplished with the aid of proteins. This is evidenced by the fact that the treatment by proteases leads to dissociation of the fibres into individual biopolymer chains and the dispersion of cells. As the egg advances in the fallopian tube, HA of extracellular matrix is destroyed and the follicular cells of *corona radiata* undergo apoptosis, the controlled death of cells, as opposed to

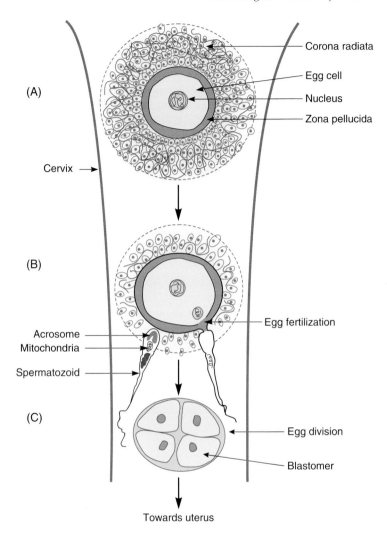

(A)

Corona radiata

Egg cell

Nucleus

Zona pellucida

Cervix →

(B)

Egg fertilization

Acrosome
Mitochondria

Spermatozoid

(C)

Egg division

Blastomer

Towards uterus

Figure 2.3   Participation of hyaluronan in fertilization of egg

cytolysis, and the uncontrollable lysis of cells. The correlation between the intensity of HA breakdown and the loss of follicular cells through apoptosis has been recently established [46]. Following polysaccharide hydrolysis, dismantling of the intercellular matrix and the disappearance of follicular cells, the thickness of the *corona radiata* coat is gradually reduced.

Simultaneously, the probability of egg fertilization is reduced. In humans, fertility of the egg cell is preserved for a period of 12–24 h after ovulation and complete uncovering of the oocyte occurs after 30 h. Hyaluronan and other glycosaminoglycans of *zona pellucida* and *corona radiata* prevent the egg from adhering to the wall of fallopian tube, thus reducing the risk of ectopic pregnancy. It has been experimentally demonstrated that retardation of HA breakdown in the intercellular matrix leads to inhibition of follicular cells apoptosis and preservation of egg fertility [46]. For the fertilization to take place, that is, introduction of spermatozoid nucleus into egg cell, the acrosome reaction must occur first [34]. This reaction begins with the contact of spermatozoid with the jelly-like coat of the egg cell. The contents of the spermatozoid acrosome are freed via exocytosis. Among hydrolytic enzymes in the acrosome, the hyaluronidase is present to break down the biopolymer.

Future studies will help us to understand the fine molecule mechanisms of participation of HA in early stages of embryogenesis. But thus far, one can assume that the products of sequential fragmentation of hyaluronan participate in the processes of fertilization, initiation of cell division, and the cleavage of zygote. These assumptions are based on a number of facts described in the literature concerning a variety of HA's functional properties as a function of its molecular weight (Figure 2.4). It is known, for example, that HA with the molecular weight of 500 000 Da and higher suppresses cell proliferation and migration. Once the molecular weight is reduced to 50 000–100 000 Da, however, suppressive activity is reversed, that is, HA starts to stimulate cell proliferation and migration. Thus, hyaluronan can perform mutually opposite functions, depending on the size of macromolecules. The medium-size fragments (25–50-disaccharides long) show immuno-stimulating, strong angiogenic activity and accelerate inflammatory processes. Oligosaccharides of smaller sizes induce expression of heat-shock proteins and act as antiapotosis factors. Some oligosaccharides of small molecular weight are, apparently, endogenous and signal factors of cellular stress, while others trigger various signal pathways of regulation. Thus, HA with the molecular weight of 1300–4500 Da (3–10 disaccharide units) initiates proliferation of endothelial cells. On the contrary, the fraction with the molecular weight in the range of 1300–7200 Da (3–16 disaccharide units) definitely inhibits proliferation of normal epithelial cells but does not suppress proliferation of normal fibroblasts and normal smooth-muscle cell *in vitro.*

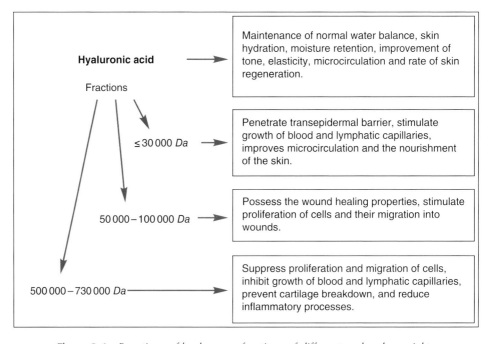

Figure 2.4   *Functions of hyaluronan fractions of different molecular weights*

Hyaluronan with a molecular weight of 100000 Da stimulates the proliferation of the human fibroblasts in the collagen matrix, and the polymer with a molecular weight of 860000 Da accelerates the proliferation of corneal epithelial cells. On the contrary, a biopolymer with a molecular weight of 400000 to 1 000000 Da, inhibits the proliferation of normal endothelial cells. Polysaccharides with a molecular weight of 1 000000 Da will suppress the proliferation of the rabbit synovial cells and fibroblasts from 3T3 rats, but at 1 mg/ml was shown to stimulate proliferation of the rabbit synovial cells and fibroblasts from 3T3 rats.

Such varying results can be associated with the differences in biopolymer concentration, presence of various impurities, and differences in the HA receptor expression in different cell types, among other phenomena. It is obvious, however, that hyaluronan possesses wide spectrum and size-dependent activity, making HA fragments powerful signal molecules for regulatory systems [47]. Isolation and functional characterization of oligosaccharide fragments is an important field of research in molecular biology and medicine.

2.2.2   Hyaluronan and Other Glucosaminoglycans in Cell Division, Migration and Differentiation

The organism of any multicellular animal can be considered as a clone of cells formed from one impregnated egg cell. Therefore, all cells are genetically identical, but differ in phenotype since they interpret genetic 'instructions' in different ways, arranging themselves with the time and the environment in the multicellular organism (i.e. they differentiate). After fertilization, the development of vertebrates can be divided into three phases. The first phase involves division (cleavage) of the impregnated egg cell (zygote) into multiple cells via mitosis and the formation of blastula, a single-layer and closed ball-shaped structure. These cells form an epithelium-like layer and as cells of all epitheliums they possess polarity. For instance, their surfaces (external, internal and lateral) are distinguished by their chemical composition and their functions, which is important for coordination of their further behaviour. The cells of blastula are mutually connected by slit contacts and extracellular hyaluronan-containing matrix. The spaces between the cells are constantly enlarged, which is associated, apparently, with the active synthesis of hyaluronan and other components of extracellular matrix. Glycosaminoglycans and proteins that cover the cell membranes' surface and form the intercellular matrix between the adjacent cells are carriers of the cell-specific surface properties that determine the further fate of cells. In contrast to the sulfated glycosaminoglycans, hyaluronan is capable of accelerating the growth of cells and their proliferation.

The phase of cleavage and the formation of blastula is followed by the phase of gastrulation, which involves the layering of the germ leaves from which tissues and organs are developed in the third phase. These phases do not have clear boundaries and can overlap. The processes of proliferation, migration and sequential differentiation of cells are intensive at all three stages. It is believed that hyaluronan plays an active role in all these processes, since accumulation of HA coincides with the periods of the migration of cells in tissues. Being the main component of the pericellular (near-cellular) layer formed on the surface of the proliferating and migrating cells HA facilitates the separation of cells. Furthermore, it can form channels of the different degree of hydration and a less dense matrix to facilitate migration of cells. As many HA-specific receptors called hyaldherins were discovered, it is possible to suggest that HA influences the migration of the cells by direct interaction with these receptors (see Table 2.2).

Table 2.2   *Hyaldherins, proteins specific for HA binding [48]*

Cellular Proteins		Extracellular Proteins	
Wandering Proteins	*Transmembrane Proteins*	*Extracellular Matrix Proteins*	*Soluble Proteins*
cdc37	Family CD44	Versican (VCAN)/	alpha-trypsin inhibitor
p68 (gclqR)		hyaluronectin	
RHAMM*		Connected protein	
HBP**		Aggrecan	
IHABP***		Neurocan	
		Fibrinogen	
		TSG-6****	
		Collagen VI	

*Receptor for hyaluronic acid-mediated motility
**Hepatocyte-binding protein
***Intracellular hyaluronic acid-binding protein
****Tumour necrosis factor – stimulating gene 6

It is considered as well-established fact that HA stimulates the proliferation of human fibroblasts through the collagen of the extracellular matrix [48] with which it is associated [49]. The production of HA does not require the active synthesis of the protein, since the already-formed synthetic apparatus is localized on the inner surface of the cytoplasmic membrane where it becomes insensitive to the action of the inhibitors of the protein synthesis and extracellular proteases. The chains of HA pierce the cell membrane that immediately merges into the extracellular matrix. Cleavage of these chains by extracellular hyaluronidases stimulates synthesis of hyaluronan. It is possible that the polysaccharide plays an important role in the regulation of the protein component synthesis of the intercellular matrix, either accelerating or slowing down its formation. The molecular weight-dependent features of the extracellular matrix and extremely wide range of conformations adopted by the polysaccharide macromolecule in it can be responsible for the control of division, migration or the differentiation of cells [49,50]. Recent communications suggest the presence of intracellular and intranuclear HA in dividing cells [51]. By methods of confocal microscopy, it was detected that in fluorescin-treated permeable fibroblasts, HA is localized in nuclear structures and chromatospherites. After stimulation of the 3T3 cells (mouse fibroblasts) with serum, a considerable increase in the cytoplasm content of labelled HA was detected, especially during the prophase and the early metaphase of mitosis. In the non-stimulated cells the labelled biopolymer is localized only in the vicinity of nucleus, especially in the chromatospherite and in regions of heterochromatin around the nuclear periphery. In the stimulated mitotic cells, the labelled HA was found to fill the entire nuclear space surrounding the chromosomes during the metaphase, and then cover the region between the separated chromosomes during the anaphase. Fluorescence from labelled HA also persists during the telophase. In the late telophase, a small quantity of hyaluronan can be detected in the vicinity of thin diaphragm 'bridges' which connect two divided cells.

After the stimulation of cells with embryonic bovine serum or thrombocytic growth factor, the labelled HA (after addition into the cultural medium) was detected in the phials and cell cytoplasm. Distribution of HA in the cytoplasm occurs compatibly with the rough endoplasmic reticulum and endosomal phials. This raises the question about location: where do endosomal phials transport HA? Into an endoplasmic network of reticulum or into the lysosomes?

It is assumed that labelled HA appears in the cell as a result of internalization via endocytosis. Inside the cell, HA can be bound to intracellular factors and participate in the regulation of the transcription of genes, a cellular cycle, or be cleaved in the lysosomes where it would be further utilized in the synthesis of hyaluronan as a precursor. Hence, there is a cycle that includes the synthesis of hyaluronan on the cytoplasmic membrane, followed by its extradition in the intercellular matrix, as well as the reverse process of transport of HA back into the cell via endocytosis. In this cycle, the partial and complete depolymerization of the polysaccharide occurs. One can assume that the controlled fragmentation of HA results in the formation of functionally active fragments that can fulfil regulator functions of gene activity control and cell proliferation. The intensity of the processes of biopolymer synthesis and breakdown during embryogenesis is especially high. It is believed that an increased content of hyaluronan in the intercellular matrix can facilitate separation of cells during division and migration. Upon microscopy observation, the labelled HA in the intercellular matrix and on the surface of the cell appears as an intricate downy network. The treatment with streptococcus hyaluronidase leads to the disappearance of labelling both on the surfaces of the cell and the pericellular matrix. Both synthesis and secretion of HA are known to increase during the cellular proliferation. In the cellular culture of non-stimulated cells, only 25% of cells showed the presence of pericellular matrix. After cell stimulation by thrombocytic growth factor, the number of cells covered with the coat of HA increased up to 70%. Almost all myotic cells were covered with hyaluronan coat. At the same time, the polysaccharide coat was thicker in the stimulated cells compared to the control cells. The formation of the pericellular coat occurred when the cells were separated from the surface of substrates during the mitotic rounding, which facilitated the slip of cell along the substrate. Thus, the amount of intracellular and extracellular HA increases in a coordinated manner during the mitotic multiplication of cells. Such a phenomenon was also observed for some other types of cells, including epithelial, endothelial and tumour. Nevertheless, the source and functions of intracellular HA are not yet fully understood.

After the cells of blastula have formed an epithelial layer, the layer of extracellular matrix is synthesized around the single-layer spherical blastula. This structure is a starting point for gastrulation - the active migration of cells and displacement of cellular masses. The layer of extracellular matrix contributes to these processes. The layering of three germ layers (i.e. ectoderm, entoderm and mesoderm) and the formation of a three-layered embryo is the eventual result of gastrulation. The cells of these layers are already determined to shape different tissues and organs. The processes of active proliferation, migration and genetic differentiation of cells also accompany the processes of histogenesis and organogenesis [52–55]. Gastrulation, histogenesis and organogenesis all feature an increased content of hyaluronan. A large quantity of biopolymers during embryonic growth is concentrated in the amniotic liquid (up to 20 mg/ml) and in the umbilical cord (up to 4 mg/ml). The layering of cartilaginous chords at this period of development also requires availability of HA, chondroitin sulfate, dermatansulfate and proteins; the building blocks of supramolecular complex of aggrecan, which is the basic component of cartilaginous tissue. Therefore, it may be assumed that the genes encoding the synthesis of HA and other glycosaminoglycans belong to the group of ancient genes that once shaped the regulatory type of embryonic growth [3], the active expression of which is especially necessary during embryogenesis.

To date, the molecular mechanisms of the hyaluronan as its functions in embryogenesis processes are poorly studied and are the important subjects for future research. Only certain details are known about the functioning of hyaluronan at the molecular level in cell culture models of regeneration, repair and recovery of the damaged animal organs. It is believed that the processes of recovery of damages of cells, tissue and organs share certain similarities with the processes of embryonic growth. Let us review some experimental results. Using a method of polymerase chain reaction, the activity of the expression (synthesis) of mRNA (matrix RNA) for all three HAS_1, HAS_2 and HAS_3 enzymes, as well as the dependence of cell migration rate on the concentration of hyaluronan in the damaged monolayer, was determined in the monolayer of the peritoneal mesothelial human cells [56]. The results showed that an increase in the activity of the gene (increase in the expression of mRNA) for HAS_2 occurs 6 h after damaging the monolayer of cells and a maximum of gene activity with a 15-fold increase in the quantity of mRNA was reached after 12–24 h. After this, the expression of mRNA of HAS_2 decreased. After 48 h it had almost reached the level of control intact cells (Figure 2.5). The content of mRNA of HAS_3 enzyme was constant in the normal cells. The level started to decrease 6 h after wounding and a six-fold decrease was detected after 24 h; the content of mRNA returned to a normal level after 96 h (Figure 2.5). The expression of mRNA of HAS_1 was not discovered in either control or damaged cultures.

It follows from the results that the activity of the three genes encoding hyaluronan synthases is regulated differentially in response to the damages. It was also established that the synthesis of hyaluronan is activated in the front edge of the wound, and that repair in the presence of HA is accelerated through increase in the rate of cell migration into the wound (Figure 2.6). It is characteristic that in both embryogenesis and monolayer of peritoneal mesothelial human cells the gene of HAS_2 is activated in response to damage [32,56] (Figures 2.6 and 2.7).

Another example is restoration of bone damage. The surface of a bone is covered with a connective tissue (periosteum) with a small number of cells capable to form new cartilaginous tissue. Upon a damage, say a bone fracture, these cells start 'repairing' by reproducing the initial embryonic processes of the bone formation from the cartilage models. First, cartilage is produced to fill the break of a bone. Then the cartilaginous tissue is replaced by bone tissue with the aid of the osteoblasts and the osteoclasts [45]. The age-related qualitative and quantitative composition of glycosaminoglycans in different types of cartilage tissue has been quite well studied. The characteristic property of the qualitative composition of glycosaminoglycans at the early stages of the embryogenesis of vertebral animals is the absence of keratan sulfate in all types of the cartilage models [57–60]; at this stage there is an increase in the content of chondroitin sulfates [60] and degree of glycosaminoglycan sulfation [61–63]. As embryo development progresses, the content of keratan sulfate increases and the content of heparan sulfate decreases [60]. A clear correlation is observed between the intensity of glycosaminoglycans and collagen exchange. In both cases the maximum intensity of metabolism is observed at the beginning of ontogenesis, especially during histo- and organogenesis [61,64]. Moreover, the level of the glycosaminoglycan exchange reaches

(A)

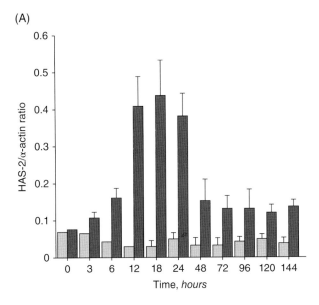

(B)

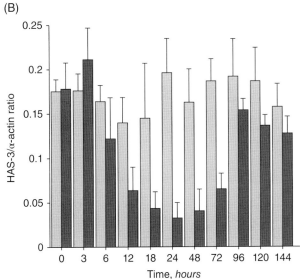

Figure 2.5   *Transcription of HAS 2, HAS and α-actin following injury of human peritoneal mesothelial cells (HPMDs). Data presented as the ratio of HAS 2/ α-actin (A) and HAS 3/ α-actin (B) for the injured cells (■), compared to control experiments (□). Reprinted by permission from Macmillan Publishers Ltd: Kidney International, from ref. [56], copyright (2000)*

a maximum earlier than the collagen exchange [63]. A similar situation is observed during regeneration of wounds of connective tissues in adult organisms. Thus, the early stages of skin wound healing are accompanied by accumulation of hexosamine [65] (Figure 2.8).

(A)

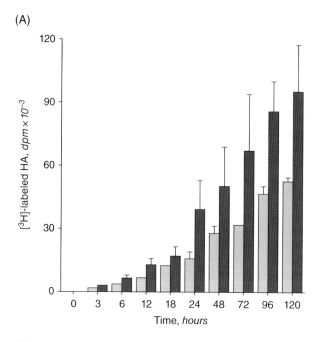

(B)

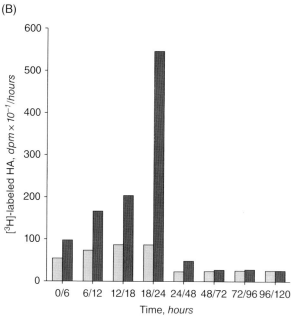

Figure 2.6 *Synthesis of hyaluronan by wounded HPMCs [56]. (A) HA accumulation at the times indicated was determined according to the amount of label incorporated into papain-digested and hyaluronidase-resistant [³H] macromolecules; (B) The control and injured cultures were pulse labelled with [³H]-glucosamine for the time indicated. The rate of HA synthesis is expressed as [³H]-HA dpm/h. Injured cells (■) are compared with control experiments (□). Reprinted by permission from Macmillan Publishers Ltd: Kidney International from ref. [56], copyright (2000)*

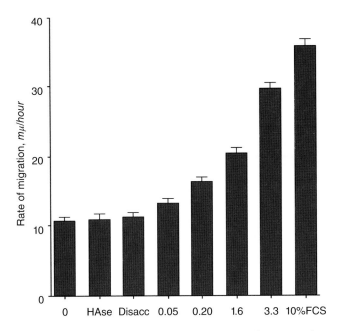

Figure 2.7 *Effect of exogenious HA on the rate of migration of HPMCs after wounding [56]. Reprinted by permission from Macmillan Publishers Ltd: Kidney International from ref. [56], copyright (2000)*

The progress of wound closure, collagen and hexozamine (as a marker for proteogly-cans) contents were studied in the normal wounds comparing with the treated by gelatin microspheres with an antimicrobial agent. It was shown that the wound closure progressed from day 0 to day 15, at the same time the maximum of hexosamine content was seen on day 3 of treated and day 9 of non-treated wounds, but the maximum of collagen content was observed on day 6 with the treated and day 12 of non-treated wounds. These observations clearly indicate that proteoglycan complexes containing glycosaminoglycans are formed in the beginning of the productive phase of growth. But the first collagen fibres appear only at the end of this phase. Consequently, the synthesis of glycosaminoglycans and proteoglycans always precedes the synthesis of collagen and the formation of scars. Most likely, glycosaminoglycans determine the scar type. The treatment accelerates the wound healing but does not change a dynamics of synthesis of proteoglycans and collagen [65]. There is also strong experimental evidence that chondroitin sulfate and sulfated glu-cosamine are capable of stimulating the synthesis of collagen as well as proteoglycans in the cartilage tissues [66–68]. Furthermore, it was shown that glycosaminoglycans at low concentrations (less than 10 mg/ml) accelerate the proliferation of chondrocytes in the cell culture, but at higher concentrations (150–200 mg/ml) suppress proliferation [65]. Such effects of glycosaminoglycans can be described as being associated with either construc-tion and/or maintenance of the intercellular matrix with characteristic properties, or by utilization of intermediate metabolites. It was experimentally demonstrated that sulfated glycosaminoglycans suppress activity of enzymes participating in the degradation of the intercellular matrix, such as hyaluronidase [65] and granulocyte elastase [69,70],

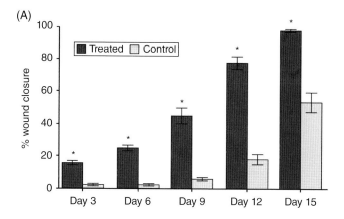

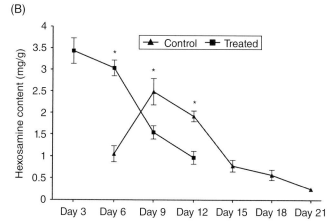

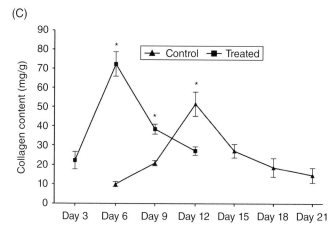

Figure 2.8 *Progress in wound healing (A), dynamic of hexosamine content (B) and collagen content (C) in skin after wounding [65]. Reproduced with permission from ref. [65]. Copyright © 2009, John Wiley & Sons, Ltd*

collagenase [71], serine proteases [72], acid cathepsins, lysosomal hydrolases [73] and other enzymes [74,75]. Interestingly, similar properties are known for monosaccharide sulfate glucosamine [76]. Such properties of glycosaminoglycans can serve as an explanation for the phenomenon of suppression of chondrocyte apoptosis *in vitro* [77].

The intermediate metabolites of glycosaminoglycans (sugar, glucosamine and sulfates) can stimulate exchange reactions and growth of the cartilage tissue [78]. Even individual metabolites manifest high physiological activity. For example, sulfated glucosamine affects the turnover and proliferation of cartilage tissue cells and chondrocytes [66,68]. A small decrease in sulfate concentration in the biological fluids from 0.3 ppm to 0.2 mm leads to a sharp reduction (up to one third) in the rate of proteoglycan synthesis in the human cartilage tissue [79,80]. It thus follows that the regulatory chains with direct and reverse feedback to control cell metabolism, division and apoptosis includes hyaluronan and sulfated glycosaminoglycans.

Cell migration is yet another phenomenon that plays an important role in the processes of embryonic development. Prior to the beginning of migration, cell types are not completely determined. In order to reach the place of their designation, cells migrate through other tissues of embryo. This is followed by differentiation of these cells in the corresponding places of destination, which is determined by mutual contacts and chemical factors of cellular membranes and extracellular matrix. The routes of migration are determined, in particular, by connective tissues [81–83].

It has already been mentioned that HA participates in the cell migration, creating a favourable gel medium in the intercellular spaces and also through the contacts with the specific protein receptors. Molecular mechanisms underlying cell migration are poorly investigated. Irrespectively of the exact mechanism beyond migration phenomenon, cells penetrate the different sections of the embryo where they are differentiated in a variety of ways to fulfil corresponding specific functions. The differentiation of cells is determined by local environment [84,85]. The extracellular matrix participates in differentiation and retention of the differentiated state of cells [86–88]. Thus, one impregnated egg cell in the process of embryogenesis gives rise to a complex multicellular organism comprising approximately 200 types of differentiated cells.

2.2.3 Hyaluronic Acid and Sulfated Glycosaminoglycans in Maintaining a Differentiated Status of Cells

Once the extracellular matrix is synthesized by cells, it helps to support their differentiated state [44,45]. The following phenomenological data serves as a basis for this conclusion. The differentiated cartilaginous cells (chondrocytes) grow and proliferate in the permeable and suited medium, eventually forming clones of the differentiated chondrocytes. They synthesize and cover themselves with large quantities of a very specific cartilaginous matrix. The cartilaginous phenotype is preserved for several generations of sub-cloned cells if they grow in the permeable medium. But if chondrocytes are grown at low density in the non-standard medium after several repeated sub-clonings, the fraction of cells that have undergone fundamental changes gradually increases. Instead of a type II collagen characteristic for the cartilage tissue, the cells start synthesizing a type I collagen typical for fibroblasts. These two types of collagen are controlled by different genes. Consequently, the switching of genes and even the possible reprogramming of the genome occurs in those cells. Under such conditions the part of chondrocytes is apparently converted into fibroblasts. Switching most likely occurs quite rapidly, since simultaneous synthesis of both types of collagen is observed only in very few cells. The mechanisms of switching are not completely known. However, some data suggests that polymeric molecules of extracellular matrix secreted by chondrocytes and fibroblasts participate in the process. Chondrocytes grow in the form of clones and form separate colonies on the surface of the solid cultural medium. The density of cartilage matrix is higher in the centre of the colony compared to that in periphery. Peripheral cells switch earlier than central cells to begin synthesizing a type I collagen. It was hence assumed that the matrix formed by chondrocytes participates in preserving a certain phenotype of chondrocytes and their differentiated status [31,45].

There are other experimental data in favour of this hypothesis. When the cartilage-specific proteoglycans (consisting of the sulfated glycosaminoglycans covalently bound with the linear protein) are added to the culture of chondrocytes, the synthesis of a cartilage-type matrix by chondrocytes is reinforced. Hyaluronan also influences the differentiation of chondrocytes, but oppositely. Fibroblasts secrete a large quantity of HA, whereas chondrocytes synthesize comparatively less. Free HA added to the culture of chondrocytes strongly suppresses the synthesis of cartilage matrix [1,44,45]. It is possible to assume that the components of extracellular matrix act as a component of the positive feedback in the regulator systems, thanks to which the synthesis of intercellular matrix becomes a self-sustaining process.

Such experimental results provide grounds to suggest that hyaluronan, sulfated glycosaminoglycans and other cell-secreted compounds of intercellular matrix act on the same cells and, therefore contribute to the retention of the genetic programs in the differentiated cells. If the intercellular matrix influences the differentiation of a cell, then the surrounding cells that secrete the compounds of the intercellular matrix should influence this process, too. Apparently, chondrocyte cells interact cooperatively and stimulate themselves to synthesize the cartilage-like matrix, while fibroblasts act as antagonists and suppress the differentiation of chondrocytes. Therefore, the differentiated state of cells is not completely autonomously preserved. Intercellular interactions can change the state of cell differentiation [46]. It can also be changed by the action of hormonal factors.

Radical changes in the cellular differentiation can occur as a result of action of the substances that change degree or nature of the DNA methylation [of 47–50]. However, these changes in the properties of the differentiated adult organism cells are rarely radical. The majority of those alterations can be considered modulations of differentiated statuses or reversible transmutations of similar cellular phenotypes [45]. Nevertheless, these modulations of differentiated cells statuses are associated with the gene switching or reprogramming of the cell genome. How can HA and other glycosaminoglycans influence genome and induce modulation of differentiated status of cells? As already mentioned, hyaluronan is synthesized by the hyaluronan synthase localized on the cytoplasmic membrane cell and 'pushed out' immediately during the synthesis into the intercellular matrix to occupy sites of specific dislocation and functioning. Therefore, direct contacts of HA with the chromosomes and genes in the nucleus are non-existent, making it possible to assume only distant interactions. These interactions are possibly realized through the binding with transmembrane receptors CD44 (Figure 2.9) or mediated by interactions with hyaldherins of the extracellular matrix, which in turn are connected to the internal cellular cytoskeleton through transmembrane proteins.

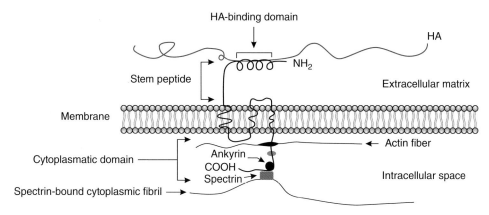

Figure 2.9   *Schematic structure of the HA-specific receptor CD44*

The exact molecular mechanisms behind the action of HA on genomes and activity of the genes through this distant signal system have thus far been insufficiently investigated. However, it is known that high-molecular weight HA (500 000 Da and higher) suppresses cell proliferation (Figure 2.4). As the presence of hyaluronan inside the cells and its intranuclear localization has been revealed [51], it is possible that the polysaccharide macromolecule, its fragments with the specific conformations, or complexes of HA with other compounds, directly affects the activity and switching of genes. In this case, a question arises around the ways HA is transported into the nucleus.

Thus, in the period of embryonic growth, cells proliferate and differentiate into different cell types. And although differentiated states are, as a rule, quite stable, some cell types are subjected to limited reversible changes or so-called modulations. For example, chondrocytes can be converted into fibroblasts under non-optimal conditions and return to the initial state as optimum conditions become available. Small reversible changes in the differentiated state are possible in many other cell types. The type of extracellular matrix synthesized by cells along with the high-molecular hyaluronan constituent help maintain the differentiated state of the adult organism cells. However, preservation of the differentiated cells in the majority of tissues of vertebral animals is based not on their longevity, but occurs through replacement of the old cells by new ones. In order to accomplish this, the differentiated cells should undergo occasional cellular cycle of division and proliferation. These processes also form the basis of physiological regeneration and recovery of tissues and organs after damage.

2.2.4   Hyaluronan and Induction of Cellular Cycles for Differentiated Cells

The integrity of a cell is based not on the strength of the materials it is built from, but on the coordination of the processes of breakdown and synthesis [89–94]. Cells are capable not only of continuous self-renewal but also of self-reproduction. The majority of the populations of the differentiated cells in vertebrates are subjected to self-reproduction. When cells are 'worn out' in biological terms and perish, they are replaced by new cells formed as a result of mitotic division of the differentiated cells. The differentiated state and the mitotic cycle of division are alternative states of cells that are associated with de-differentiation, that is, the reprogramming of the genome and cardinal reorganization of cellular structures. In the differentiated state, the group of genes responsible for the synthesis of proteins and other biopolymers secreted to fulfil specific tissue functions is active. With the initiation of the cellular cycle, this 'export' group of genes is turned off and the group of genes responsible for intracellular syntheses becomes activated. This is why the state of differentiation and the state of mitotic division to a certain degree compete with each other.

A hypothesis regarding the competitory relations of proliferation and differentiation has already been formulated and substantiated by Brodsky et al. in 1981. [95]. Brodsky claims competition occurs because of the limited metabolic possibilities within a cell that results in competition of cellular functions. A differentiated cell not only synthesizes a unique composition of proteins, polysaccharides and other biopolymers for fulfilling tissue-specific functions, but also biopolymers for internal cellular renewal. When the cell enters the cell cycle, however, the genome is reprogrammed to the synthesis of chromosomal proteins, proteins for DNA replication and an apparatus required for division and advancement of a cell through the phases of cellular cycle. In the premitotic phase, the mass of the entire cell doubles. The cellular syntheses of different biological purposes are ensured by common uniform metabolic reactions that involve macroergic compounds such as ATP, UTP and so on, precursors, reducing molecules and so on. In a single differentiated hepatocyte cell, for example, an hour-long protein synthesis period is observed that includes syntheses of proteins for secretion as well as intracellular proteins [96]. Under non-optimal conditions, when the scarcity of metabolites and synthetic possibilities is encountered, the cell enters a cellular cycle. The supply of metabolites depends, in particular, on the ratio of the cell surface to its volume. Collision is cell-shape dependent, since in round cells the volume increases more rapidly ($V = 4/3\pi r^3$) than in flatter surface cells ($S = 4\pi r^2$). The lipid composition and structural integrity of a cell membrane are also important, since these factors determine activity of the transport systems of a cell, particularly distribution of the basic phases of gel, liquid crystal, ordered liquid crystal or raft. In functional terms, the structure of the intracellular matrix and the physico-chemical state of cytosol (gel-sol transitions) are also important. All these factors determine synthetic possibilities and steady states of cells. For example, a limiting factor of the retention of the phenotype (a differentiated state) of cartilage cells (chondrocytes) is, in particular, the concentration of uridine triphosphate (UTP) in the cultural medium [96]. Formation of the precursors of glycosaminoglycans synthesis, such as UDP-glucuronic acid, UDP-iduronic acid, UDP-acetyl glucosamine and UDP-acetyl galactosamine largely depends on the concentration of UDP. Under the optimum cultivation conditions, chondrocytes perform various functions. For instance, they synthesize chondroitin sulfate for the extracellular matrix and in the presence of proliferation activators some cells divide. When the amount of UDP in the cultural medium is limited, cells cease chondroitin sulfate synthesis and begin to divide. In this case the phenotype of cells has changed. There are examples of the accelerated cellular differentiation as a consequence of retarded proliferation of cells. When chondrocytes are grown for many generations on the not-entirely suitable medium and lack the ability to produce cartilage matrix, as the phenotype changes to a fibroblastic one they are further transferred into the permeable medium where the cells regain the ability to synthesize glycosaminoglycans and cartilage matrix [86–88,97].

Multiple models are available for investigating the mechanisms of cellular transitions from the differentiated state into the cellular cycle: the cell cultures on solid and liquid nutrient medium, the damage of the cell monolayer, the wounding of the skin and other connective tissues, hepatectomy of the liver and so on. A number of activators and inhibitors of cellular proliferation are found [98]. The differentiated cells preserve mitosis capability in the tissue where they reside for long time where there is no stem cell reserve, and where the reproduction is the only source of cell mass growth. Such cells are heart myocytes, hepatocytes, endothelial cells of blood vessels and so on. The most popular model for investigating the mechanisms of the transition of differentiated cells into proliferative ones is via the liver hepatocytes in adult organisms. This is because the population of hepatocytes is functionally homogenous and all hepatocytes perform one and the same spectrum of functions. Under normal conditions, hepatocytes renew themselves at a low but strictly controlled rate, that is, through physiological regeneration. Most importantly, through experimentation it was discovered that after removal of 2/3 of a liver from an animal, the remainder of the hepatocyte population enters the cellular cycle to regenerate the initial size of the organ [95,97–99]. Besides the surgical removal of the liver, the proliferation of hepatocytes can be induced by carbon tetrachloride, ionizing and nonionizing radiation, ultrasound, short-term heating, mechanical damages, treatment of cells with proteases, detergents, inhibitors of sodium and potassium ATPase [100], colchicine and vinblastine [101], with low doses of carcinogens [102], by changing the ionic medium [103], by reversible inhibition of protein synthesis [104] and so forth. Many of these factors disrupt or destroy hyaluronan, proteoglycans and other structures of the cellular matrix. It is possible that these destruction processes act as a trigger prompting the differentiated cells to enter the cellular cycle. In the least it is known that the enhancement of the synthesis and secretion of hyaluronan, followed by the transport of HA back into cell and nucleus via endocytosis, are among the first responses to such stimuli. Stimulation of cells to the division is accompanied by accelerated degradation of both extracellular and endosomal (intracellular) HA. In other words, turnover of hyaluronan is accelerated [51].

The products of intracellular degradation of HA can play a regulatory role in the processes of cellular proliferation. The stimulatory properties are attributed to the polysaccharide with a molecular weight of 50 000–100 000 Da (Figure 2.4). It has recently been shown that oligomeric HA of a certain size (oHA) stimulates rapid proliferation of endothelial cells [105]. In the latter case, oHA induces activation of the genes of early response, in particular the *ras* gene, as well as the protein kinase C and MAP kinase. The possibilities of signal transduction from oligomeric HA to nuclei genes should also be considered. The pathways and the mechanisms of the signal transfer are examined next.

Upon mechanical damage of the monolayer of human mesothelial cells, the gene encoding hyaluronan synthase-2 is activated, whereas on the contrary, the gene responsible for hyaluronan synthase-3 is suppressed. This conclusion is based on experimental data obtained using the PCR method that showed a rapid 15-fold and six-fold increase in expression of hyaluronan synthase-2 mRNA and hyaluronan synthase-3 mRNA, respectively [56]. The gene for hyaluronan synthase-2 apparently relates to the group of the genes of early response that are silent in the differentiated cells but become first activated genes in response to the proliferation stimuli. The following genes belong to this group of genes: *c-myc* (it encodes the protein nucleoside diphosphate kinase B), *c-fos*, *c-ras*, *c-jun*, and P53. Earlier they were known as proto-oncogens, since their activation occurs when a normal cell is transformed into a cancer cell. The genes controlling synthesis of proteins in response to a sharp increase in the temperature, that is heat stress proteins or HSP, also belong to the same group of genes. HSPs, as the name suggests, are called to reduce the cellular damages under unfavourable actions such as high temperature and so on. Heat stress protein HSP70, for example, suppresses apoptosis (genetically programmed cell death) by preventing the production of caspase 3. The family of the caspase genes encodes the group of proteases that participate in apoptosis through cascade activation of the enzymes, ultimately leading to the death of a cell. It was shown on the model of arthritis that intra-articular therapy with administration of HA with a molecular weight of 840 000 Da (HA84) leads to suppression apoptosis of synovial cells and regulation of the HSP72 expression. In the same experiment, the kinetics of HA84 degradation in the synovial tissue was studied. It was established that the oligomeric products of HA84 degradation by regulating expression of HSP72 can reduce the damages of cells. The oligomeric fragments of different molecular weight were then prepared and studied for their HSP72 expression-stimulating and apoptosis-inhibitory activity. It turned out that all these functions are ascribed to the tetrasaccharide fragment of hyaluronan.

Other products of HA depolymerization, that is di-, hexa-, deca- and dodecasaccharides do not possess such activities. Importantly, the tetrasaccharide obtained from the sulfated glycosaminoglycan keratan sulfate did not show such properties, either. The tetrasaccharide of HA suppresses cell death and regulates the expression of HSP72 both at the mRNA transcription (activity of gene) and translation (activity of the synthesis of protein) levels but only in the cells subjected to hyperthermal stress (42–43 °C for 20 min). For the intact cells not subjected to shock temperature, the tetrasaccharide of hyaluronan affected neither the level of the HSP72 expression nor the viability of the cells. The molecular mechanism of regulation of the HSP72 expression under conditions of stress was also established. It appears that tetrasaccharides activate the heat stress factor HSF_1. As is known, upon activation the protein HSF_1 is phosphorylated and further transported from the cytoplasm into the nucleus, where it binds with the DNA element of heat stress, thus activating the gene responsible for the HSP synthesis in stress situations [106].

Another population of the differentiated cells capable of renewal through mitotic division similar to the hepatocytes is the endothelium of blood vessels. The endothelial cells form a single layer lining all blood vessels and regulate the exchange of substances between blood and the surrounding tissue. The extracellular matrix of endothelium contains a significant amount of hyaluronan. New blood vessels are developed in the form of outgrowths of endothelial cells from the walls of the small capillaries. These cells are capable of forming hollow capillary structures even in the cell culture. Damaged tissues, and some tumours in adult organisms, stimulate the formation of new capillaries (angiogenesis) by the adjacent endothelial cells with the aid of the special activators. Among such molecules is HA of low molecular weight [107].

It has been shown that the products of HA depolymerization that are 4–20 disaccharide units in length stimulate proliferation of endothelial cells, their migration and the initial stages of the capillary shaping *in vivo* and *in vitro* [107]. While low molecular weight HA stimulates angiogenesis, high-molecular weight biopolymer, on the other hand, inhibits the growth of new blood vessels. It is known that the growth of a tumour rapidly ceases if the tissue is not able to induce proliferation of endothelial cells and penetration of blood vessels into the tumour. The solid tumour growing in the form of dense mass feeds only due to the diffusion of nutrients from the periphery, and therefore it cannot grow to be more than several millimetres in size. It is known that high molecular weight HA impedes the migration of normal and tumour cells. Tumour cells capable of invasion and formation of metastases synthesize and secrete hyaluronidases into the extracellular matrix to cleave hyaluronan. Further studies of the mechanisms of the influence of hyaluronan on the tumour growth and the metastatic spreading and the angiogenesis in tumour and wound tissues will possibly lead to new technologies of treatment and prophylaxis of tumour processes.

Thus, the initiation of the cellular cycle in the differentiated cells is accompanied by several processes including enhancement of the synthesis and secretion of HA and its transport into the nucleus [51], acceleration of the transport of non-histone proteins from cytosol into the nucleus [108], activation of nuclear polypeptide synthesis [109], sequential changes in the macrostructure of nucleolus and non-nucleolus chromatin [110] and activation of the synthesis of nuclear proteins and DNA [104,111]. Undoubtedly, all of these processes are interlinked and represent the algorithm of the chromatin re-structuring [112] and gene switching. As it was shown in the hepatocyte model [113], the genes of early response c-myc, c-fos, c-ras, c-jun and P53 [114,115] that were once silent in differentiated cells have now become rapidly activated, but the gene encoding synthesis of albumin is inactivated [116]. Thus, the switching 'the genes of differentiation' to 'the genes of cellular cycle' occurs.

Remarkably, the highest concentration of hyaluronan in cells is found in the chromatospherite in the vicinity of the nucleus membrane [51]. Specifically, chromatin in the vicinity of a nuclear membrane is composed of the most condensed form of inactive chromatin with chromosomes that are connected with the internal surface of the nuclear membrane. A high degree of chromosome organization is preserved in the interphase nucleus of differentiated cells [117]. The centromeres of chromosomes are grouped and associated with the nuclear membrane on one side of the nucleus, whereas telomeres (end-sections of chromosomes) are connected with the membrane on another side. (Figure 2.10).

The region of 10 different chromosomes containing ribosomal genes that encode the synthesis of ribosomal RNA is assembled in the nucleolus structure. The following cytological re-arrangement timed to the different stages of cellular cycle occurs in the nucleus. Initially, the 'separation' of the nucleus occurs through the lysis of the endoplasmic reticulum membranes near the nuclear membrane. Then the 'separation' of the chromosomes occurs, which requires the dismantling of the nuclear membrane and nucleolus. The separation of chromosomes at the metaphase and anaphase stages follows. The highest content of HA is noticeable specifically at these stages. The biopolymer fills the entire region surrounding chromosomes during the metaphase and the separations of chromosomes during the anaphase [51]. The role of HA in the nuclear processes is associated probably not only with its high viscosity, but also with the possibility of self-organization into the liquid-crystal ordered structures (see Chapter 4). The biopolymer apparently creates the medium in which the mechano-chemical reactions of the interphase chromatin are restructured [108–115] and the replications and recombinations of chromosomes [117] and their separation into daughter cells proceed.

At the same time, as was mentioned [106], some hyaluronan oligosaccharides can participate in the gene switching at the early stages of the cellular cycle initiation, that is, can fulfil regulatory functions. Thus, in all likelihood, nuclear HA participates in the structural

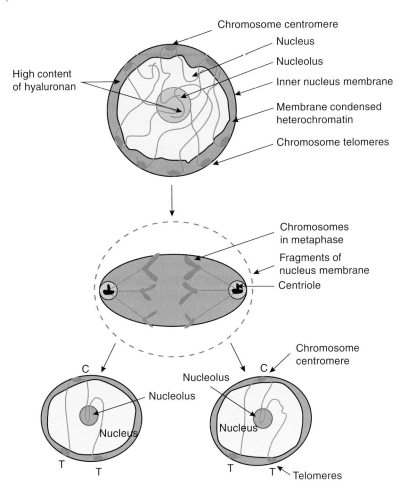

Figure 2.10 *Hyaluronan in interphase nucleus of cell (I) and cellular cycle (II, III). The regions with the highest concentration of HA are shown in red*

organization of chromosomes in the interphase nucleus of the differentiated cell, helps to dismantle chromosomes as they enter the cellular cycle (as well as their separation), and shapes the chromatin structures in the nuclei of daughter cells during their differentiation (Figure 2.10). The entering of differentiated cells into the cellular cycle is associated with enhanced hyaluronan turnover (Figure 2.11).

This turnover includes the stages of the hyaluronan synthase-2 gene activation for the transcriptions of the enzyme's mRNA, enzyme incorporation into cytoplasmic membrane followed by the synthesis and direct translocation of HA into intercellular space, further transport from the intercellular space into the nucleus and fulfilment of the functions required for reconstruction, motion and organization of chromosomes in the nucleus. There is not yet sufficient experimental data to surely confirm this common picture, but from the data available it is possible to suggest such a description. This conclusion provides directions for further experimental research.

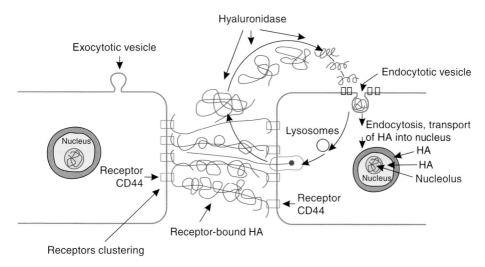

Figure 2.11 *Turnover of hyaluronan*

Not all differentiated cells in adult organisms are capable of reproducing themselves by entering the cellular cycle. This depends on which stage the cellular cycle has been stopped at prior to differentiation. If the cells are stopped at the restriction point, that is, point R, the cells in the differentiated state remain normal and viable for long time, even under conditions of nutrient deficiency. The cells that cease to divide at the 'safe point' R are so-called cells in the phase G0 of the cellular cycle. When the cells are stopped at other moments of the cellular cycle, they usually perish if they are doomed to starvation [34]. Any non-optimal conditions, including starvation, that provoke the differentiated cells to enter the de-differentiation and cellular cycle, will cause reduction in the rate of the synthesis of proteins. On the basis of this principle, a useful model was created for investigating the mechanisms of the cellular cycle initiation in the differentiated hepatocytes [104,113]. In essence, in this model, the temporarily impactful yet reversible inhibition of protein synthesis by cycloheximide antibiotic prompts some differentiated liver cells to enter cellular cycle [104], while prompting others to undergo apoptosis [118]. It is possible that these differences in how a strategic choice of response is realized are associated with the differences in the points at which the division of cells was stopped.

There are certain cells that, after being formed in a necessary quantity in embryo, remain unchanged in the course of the entire organism's life. Such cells never divide and cannot be replaced in the case of their loss. These are called permanently differentiated cells. These are almost all varieties of nerve cells, heart muscle cells, photoreceptors and cells of the crystalline lens of an eye in mammals. The macromolecules of proteins and hyaluronan are renewable in the permanent cells. Reliable evidence indicating that HA participates in the maintenance of the permanently differentiated cells is not yet available.

The third type of differentiation is found in terminally differentiated cells. In a number of tissues, the final state of cell differentiation is incompatible with the cell division. This is because the nuclei of cells can be either destroyed, that is, the cell of the outer layers of the skin epidermis, for example, or pushed out of the cells, as happens during 'ripening' of erythrocytes in mammals. Realization of mitosis and cytokinesis can be prevented if the cytoplasm is filled with myofibrils (muscle cells) or by other biochemical factors. In such tissues the need for newly differentiated cells is constant. The reproduction in these cases occurs due to stem cells. Stem cells are able to divide unlimitedly and give differentiated descendants. Stem cells themselves are not differentiated, but they are programmed to produce either one form of the differentiated cells (unipotent stem cells) or several forms (pluripotent stem cells). The model for the formation of skin epidermal cells from stem cells and their behaviour at different stages of differentiation up to the terminal one is most studied. Skin epidermis contains several well-distinguished cell layers. The deepest layer consists of basal cells located close to the basal lamina. The basal lamina separates the skin epidermis from the derma beneath. The membrane is a non-cellular structure composed mainly of collagen type IV fibres. It works as a barrier for the migration of fibroblasts, macrophages, and other cells from the extracellular matrix of derma into the extracellular matrix of epidermis. Since the blood and venous capillaries are not present in the epidermis, the membrane is permeable and transports the nutrients. Several layers of large and flat spinous cells are located above the basal cells. Layers of granular cells are located above. Closer to the surface, the tissue is composed of the organelle-free and hexagonal cells that contain only protein keratin. The basal cells adjacent to the basal lamina are unipotent stem cells and constantly preserve a capability for division. After division the cell associated with the basal lamina remains the stem cell, while another cell forms the spinous layer, the first stage of the moving upward 'escalator.' In the spinous layer, the cell irreversibly loses capability for division and starts differentiation. And, as it advances upward in the skin, the cell synthesizes different types of keratin encoded by different genes. This family of genes, similar to the family of genes for hyaluronan synthase, appears in evolution as a result of duplications and mutations of one ancestor gene [119,120].

As the cell passes from the lower 'floors' to the upper layers, it expresses different groups of genes in an entire collection of homologous keratin genes. The time from the moment of stem cell division to its desquamation from the skin surface in a human is 2–4 weeks. It is assumed that stem cells preserve capability for division owing to the contact with the basal lamina [78]. As soon as the contact is lost, the cell undergoes successive stages of differentiation, eventually leading to terminal differentiation. The basal lamina is in contact with the dermal intercellular matrix rich in HA actively synthesized by dermal fibroblasts. Apparently, there should be yet-to-be-discovered specificity of the matrix composed of collagen IV, hyaluronan and other compounds that supports 'the immortality' of stem cells in epidermis. Hyaluronan is present in the intercellular matrix of different epidermal layers (about 0.1 mg/g), but it is unknown whether HA is transported from the derma or synthesized by epidermal cells and what role it plays at the different stages of the differentiation of the epidermal cells.

In summary, the majority of the differentiated cell populations in adult vertebrates are subjected to reproduction by different ways. Most frequently this occurs by entering the cellular cycle. The division of cells is initiated by many physical and chemical factors that disrupt cell homeostasis. The common consequence of the unfavourable disrupting actions becomes, as a rule, inhibition of the protein synthesis, the most complex process with respect to number of participating components, structural organization, and high energetic cost. Unfavourable actions limit the metabolic possibilities of a cell and, first of all, possibilities for synthesis of proteins. Under these conditions, the cellular functions start to compete. The synthesis by polyribosome in endoplasmic reticulum of the 'export' proteins intended for fulfilling the differentiated functions is slowed down or entirely ceased. The cell starts synthesizing intracellular proteins for its own survival or entering the cellular cycle. This hypothesis is supported by experimental data regarding the transfer of the differentiated hepatocytes into the cellular cycle as a result of direct reversible inhibition of the protein synthesis by cycloheximide, a specific inhibitor [108–115]. This antibiotic does not disrupt the structures of the protein synthesis apparatus, but rather competitorily binds it to the transferase centre of the ribosomes and temporarily stops the machine of protein synthesis in eukaryotic cells [113]. At this moment, the algorithm of chromatin reconstruction is put in motion to turn off the group of the genes that are active in the differentiated cells and specific encoding functions and turn on the group of 'early genes' that starts the cellular cycle [113]. Among these early genes there is the gene encoding hyaluronan synthase-2. Even under the conditions of the complete blockade of protein synthesis, hyaluronan can still be synthesized by the existing enzyme, and the synthesis of precursors is even activated in the absence of the competition for the macroergic compounds (ATF, UTF), thereby reducing molecules and initial sugars as experimentally observed.

The action of unfavourable chemical and physical factors leads to occasional spontaneous mutations. The average frequency of such mutations, according to general estimates, is 10^{-6} mutations per one gene per one division cycle. In the human organism about 10^{16} divisions take place in the course of its life. Each gene during the life of the individual can undergo mutations approximately 10^{10} times. Therefore, dividing differentiated cells face numerous 'options': (1) to finish the cellular cycle normally and pass into the initial differentiated state, (2) to pass into the new steady state of the differentiated cell, (3) to adopt the immortalized state, that is, the state that occurs when the system regulating the division-to-differentiation transition is disrupted and the cell constantly divides or (4) transformation into cancer cells. When the regulatory systems are seriously damaged, the cell undergoes (5) controlled cytolytic chain process, apoptosis or (6) uncontrollable cytolysis. The two last options will lead to death of the cell.

Controlled destruction of cells in apoptosis is important at the last stage of the healing of wounds. Apoptosis apparently regulates the replacement of the temporary granulation tissue by the final scar tissue. Several types of cells, for example, fibroblasts, endothelial cells, myofibroblasts and inflammatory cells undergo apoptosis. Apoptosis is thoroughly regulated. When apoptosis is slowed down or stopped, fibroblasts continue to proliferate creating keloid tissue. Interestingly, this process involves the genes of early response, that is, c-myc, P53, which are activated with the initiation of the cellular cycle. The gene P53 controls the normal cellular cycle and the return of cells into their differentiated state. Mutations in this gene, as is known, can cause development of tumours in humans. Point mutations in the gene P53 are also found in the fibroblasts of keloid scars. In the normal skin fibroblasts of the same individuals, such mutations were not detected [121].

Thus, in the dynamic equilibrium, the adult organism consisting of 75 trillion cells grouped into 200 histological types is supported by two opposite processes: cell proliferation and apoptosis. Hyaluronan executes a striking set of functions in these processes. Being localized in the intercellular matrix, HA is forced to 'accept the fire' coming from the mutagenic influence of external physical and chemical actions as well as internal factors, in particular the omnipresent free radicals, upon itself. Thus, hyaluronan and other macromolecules reduce the mutagenic background and frequency of mutations, thereby decreasing the probability of different diseases.

On the other hand, hyaluronan can contribute to the development of pathology. Mutant cancer cells, for example, frequently have a selective advantage over normal cells. They are capable of replicating limitlessly as they 'ignore' the Hayflick principle and penetrate into the intercellular spaces of normal cells, that is, to grow invasively and uncontrollably and spread metastastically. It is shown that the extracellular matrix of cancerous cells is rich in HA and hyaluronidases that apparently participate in the metastastic spreading [122]. On one hand, tumour cells use hyaluronidase as a 'molecular saboteur', causing depolymerization of HA so that degradation products can further facilitate the propagation of a tumour by causing angiogenesis; that is, the germination of blood vessels into the tumour. On the other hand, the high molecular weight HA of the extracellular matrix prevents replication, migration and metastastic spreading of cells, and even strengthens the activity of anti-cancerous agents. Furthermore, the cells of many tissues and organs are densely encapsulated by connective tissue, and their extracellular matrix contains significant amounts of hyaluronan. The cells of some types perish and are deprived of the specific factors necessary for survival as they cross the connective tissue barrier. Epithelial cells are immobilized due to selective extracellular adhesion and anchoring to the basal lamina, the boundary between epithelium and surrounding tissues compartments. Besides collagen of different types, the basal lamina contains hyaluronan. Only few types of specialized cells, that is, macrophages, lymphocytes and branched projections of neurons (dendrites) can cross this barrier under normal conditions. The role of HA is not limited to participation in adhesion and cell and tissue compartmentalization [123–125]. The biopolymer is also involved in the maintenance of the dynamic cellular replication-apoptosis equilibrium through the signal systems of regulation [118].

The regulation systems in the superior vertebrates are ultra-stable in the cybernetic sense. They are organized according to the hierarchy principle and possess multiple-contour regulation, which allows the cells to search for the most optimum and stable states with a change in conditions. The role that hyaluronan plays in the creation of optimum and stable states and also in pathologies associated with disturbances in the signal systems, surplus or deficiency in the synthesis of HA and other glycosaminoglycans, will be examined next.

2.2.5 The Source of Hyaluronic Acid's Functional Properties and the Dynamics of its Synthesis and Degradation

In each type of human cell, tissue and organ, the normal physiological levels of substance concentration is created and maintained by the equilibrium rate of synthesis and decomposition. This general biological rule also applies to hyaluronan. The rule is involved in the support of homeostasis [126] and the timely organization of cells [127] as a biochemical oscillator, which causes un-damped oscillations of macromolecule concentrations due to the existence of the positive and negative feedback in the circuits of biochemical transformations. HA's ability to form these vibration (i.e. oscillatory) reaction modes will be discussed in the next section.

Hyaluronan is involved in many biological processes and is present in almost all body tissues of vertebrates where it plays a role in the regulation of cellular activities. It speeds up (or slows) cell division, influences migration, and is involved in the reorganization of chromatin structure and gene switching. Hyaluronan is localized in the nucleus, the cytoplasm and the intercellular matrix, where it interacts with receptors on the cell surface and intercellular matrix proteins. HA is involved in the process of a cell's adaptation to physical and chemical exposure, the process of fertilization, embryogenesis, angiogenesis, inflammation, regeneration and tumour growth. It can exhibit additive, synergistic and antagonistic properties with related sulfated polysaccharides and between hyaluronan oligosaccharides with different molecular weights.

What is the reason for such a variety of physiological and functional properties of HA, despite its relatively simple chemical structure that does not undergo any post-synthesis chemical modifications? To clarify this question, we shall examine the metabolism of HA through kinetics of synthesis and degradation *in vivo* and *in vitro*. We must investigate how the physicochemical properties and related physiological functions of hyaluronan oligosaccharides are changed and why the hyaluronan synthase family (and its corresponding genes) are needed for the synthesis of polysaccharides with a different molecular weight.

Hyaluronan has a different rate of synthesis and decomposition in different tissues. For example, HA's half-life in skin is 24–48 h. During this time, 50% of hyaluronan decomposes and the same amount is synthesized. The half-life period of skin proteoglycans, in comparison, is several days, and for mature skin collagen it can be up to several months. Skin proteins are completely regenerated in approximately 160 days.

As was pointed out above, nature created a very interesting mechanism of hyaluronan synthesis in the eukaryotic cells which is different from other glycosaminoglycans. The sulfated glycosaminoglycans are synthesized in the Golgi apparatus where they covalently bound to core proteins, forming proteoglycans. Then they are packed into vesicles and secreted into the extracellular matrix by the mechanism of exocytosis. In turn, hyaluronan is synthesized by hyaluronan synthases, which are membrane proteins integrated into the cytoplasmic cell membrane. These enzymes enlarge the HA molecule by adding glucuronic acid and N-acetylaminoglucose, one after another, to the initial polysaccharide. After elongation, the polymeric chain is transferred through the membrane into the extracellular matrix. HA can be transported into the cell nucleus from the extracellular matrix using the mechanism of endocytosis.

Such opposing directions of the synthesis and transportation of HA and sulfated glycosaminoglycans must make biological sense. One of the hypotheses suggests that at the initiation of the cell cycle by the aggressive environmental factors, the cell interrupts the export syntheses, including synthesis of proteoglycans for the extracellular matrix. During the preparation for division, many intracellular structures that synthesize core proteins for proteoglycans, including intracellular membranes, undergo partial or complete decomposition. At the same time, the cytoplasmic membrane and synthesis of hyaluronan by membrane-connected hyaluronan synthases are kept intact. Apparently they are more resistant toward different perturbations, because in such extreme conditions it is necessary to synthesize HA at an increasing rate. This was shown in numerous experiments. The increasing level of HA precedes or is coincident with the periods of initiation of the cell cycle, cell division and migration, inflammation, regeneration, angiogenesis or tumour growth.

Contrary to sulfated glycosaminoglycans, synthesis of hyaluronan does not require active synthesis of the protein. Pre-existing hyaluronan synthases as transmembrane proteins contain cytoplasmic domain in which several sites – targets for phosphorylation by protein-kinases C and A – exist. The compounds, which increase the activity of the protein-kinase C (e.g. phorbol esters) and protein-kinase A, also accelerate the synthesis of hyaluronan. Thus, the regulation mechanism of the biopolymer content by phosphorylation-dephosphorylation is created.

The preference for hyaluronan synthesis is not only due to supposed increased stability of the membrane-connected apparatus for synthesis, but also due to gene activation. As was described previously, there is a family of three genes that encode three hyaluronan synthases. During the transformation of the differentiated cells into the cell cycle, the gene for hyaluronan synthase-2 is activated. That hyaluronan synthase synthesizes a particular polysaccharide with a molecular mass up to 2 000 000 Da. The gene for hyaluronan synthase-3, which synthesizes the shorter hyaluronan chains, is active in the differentiated cells and suppressed during the transformation into the cellular cycle. These conclusions were made based on experimental data obtained by the PCR method which described the corresponding modifications in mRNA for hyaluronan synthase 1, 2 and 3 based on the model of the regeneration of the damage of the mono-layer of the mesothelial human cells [56]. The expression of mRNA for the hyaluronan synthase-1 was not found, in either the control or in the damaged cultures (up to 40 cell divisions). This enzyme performs the synthesis of HA with the highest molecular weight, which is likely necessary for preservation of the differentiated state of cells *in vivo*. Such a differential-adjustable system of gene activity on the transcription, translation and synthesis levels at different rates and different values of molecular weights indicates that hyaluronan plays a key role in selecting and determining the status of the cell and its adaptation properties.

A high rate of HA synthesis also corresponds to a high rate of its biodegradation. Hyaluronan can be degraded by the enzyme family hyaluronidases. In the human body, at least seven types of hyaluronidases have been found. Their synthesis is controlled by corresponding genes and their activities are regulated by numerous factors, including hormones (Figure 2.12).

The mechanism of action of the pituitary gland hormone vasopressin is based on the activation of hyaluronidase [2]. One of the characteristic features of hyaluronidases is formation of the oligosaccharides with different molecular weights and different functional properties. For example, HA with a molecular weight above 500 000 Da suppresses the proliferation cells, migration and angiogenesis. In contrary, depolymerization products with a molecular weight of 50 000–100 000 Da stimulate cells proliferation and migration. The oligosaccharides with molecular weight of 30 000 Da and below stimulate angiogenesis (proliferation of endothelial cells and formation of capillaries) as well as inflammation. It was noted earlier that the HA tetrasaccharide is able to activate several proteins in conditions of thermal shock and slow cells death [106]. It was also found that HA fragments – tetra- and hexasaccharides – cause immunophenotype maturation of the dendrite cells from the human monocytes [128]. The effect of the activation of dendrite cells by tetra- and hexasaccharides is highly specific, since it cannot be initiated by the HA oligosaccharides with molecular weights 80 000–200 000 and 600 000–1 000 000 Da. It is important that other glycosaminoglycans, such as chondroitin sulfate and heparan sulfate and their fragmentation products, do not have such activation effect.

It has been shown that the expression of CD44, which is the main cellular receptor of hyaluronan, is not required to reveal the effects of dendrite cell 'ripening' under the influence of small 4–16 oligosaccharides HA fragments. This conclusion is in agreement with the results of wound inflammation and healing research. Small 3–10 oligosaccharides fragments of the biopolymer accelerate neoangiogenesis during wound healing in the first 48–72 h. High

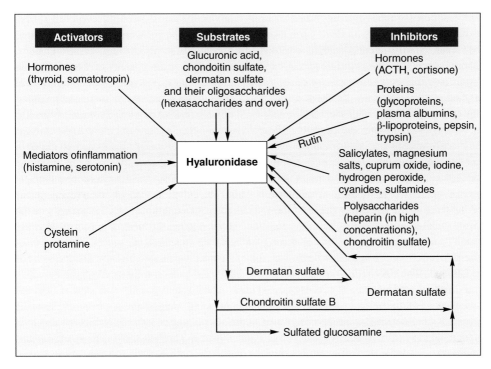

Figure 2.12   *Substrates, activators and inhibitors of hyaluronidase*

molecular weight HA does not affect angiogenesis, but at high concentration would even suppress it. It was concluded that only the polysaccharide fragments of 10–16 disaccharide units activate dendrite cells and only fragments comprising more than 6–10 oligosaccharides can bind with CD44 on the endothelial cells or keratinocytes. Other HA receptors, different from CD44 and RHAMM, do not participate in inducing the ripening of dendrite cells by the small fragments, because neither mRNA nor protein RHAMM were found in the human dendrite cells. These data suggest that the small fragments of polysaccharide created in the zones of inflammation activate the dendrite cells *in vivo*, migrating inside the inflammation tissue or outside it, thus stimulating and supporting immune response.

However, another study presented different results [129]. The oligosaccharides with molecular weight from 1300 to 4500 Da (3–10 disaccharide units) caused the proliferation of the aortic endothelial cells. The fraction with high molecular weight did not cause such an effect. The oligosaccharides with a molecular weight from 1300 to 7200 Da (3–16 disaccharide units) definitely inhibited proliferation of the normal endothelium cells but did not influence proliferation of the normal fibroblasts or normal smooth-muscle cells. HA with a molecular weight of 1 100 000 Da stimulates the proliferation of human fibroblasts but polysaccharides with a molecular weight 860 000 Da activate epithelial cells of cornea. Hyaluronan with a molecular weight ranging from 400 000 to 1 000 000 Da suppressed proliferation of the normal endothelial cells, while a polymer with a molecular weight more than 1 000 000 Da inhibited proliferation of rat 3T3 fibroblasts and rabbit synovial cells. These effects depend on concentration. In concentrations below 1 mg/ml the same

HA fraction stimulated proliferation of both types of the cells. Such differences in these experimental results, which were often observed with similar forms of HA products, could be connected to different concentrations of HA by presence of impurities (e.g. sulfated glycosaminoglycans), as well as differences in expression, quantity and sensitivity of the receptors to hyaluronan on the cytoplasmic cell membrane. Another interesting example is the model of angiogenesis. The results obtained on such a model could be interpreted quite unambiguously. It was shown that HA oligosaccharides with 4–20 disaccharide units stimulate capillary growth *in vivo,* but *in vitro* they induce proliferation of endothelial cells, their migration and the initial stages of the blood vessel formation. However, the molecular weight limits, which cause or inhibit the angiogenesis effects, were not established. Several attempts to study the direct or indirect effects on the angiogenesis were undertaken, but the precise mechanisms of such processes remain unclear. At the same time, the established phenomenology clearly indicated that the variety of the functions of HA takes place not only due to diversity of the hyaluronan synthase genes and the polymers of the different molecular weights that are produced by them, but also due to an array of their decomposition products. The variety of the HA functions strongly correlate with the variety of the decomposition products of the polysaccharide, which have different important functional properties in the living body.

Although the intensity of glycosaminoglycans exchange is significantly higher than the intensity of the proteins exchange, scientists nevertheless point out the strong relationship between the intensity of the glycosaminoglycans and collagen [65]. Such a relationship is maintained even in the case of the maximum intensity of the exchange of the glycosaminoglycans and collagen in the beginning of ontogenesis and the period of the formation of the tissue structures. The most studied effect is the exchange dynamics (synthesis – decomposition) of glycosaminoglycans and collagen in the cartilage matrix, where a large amount of the sulfated glycosaminoglycan chains are covalently bonded with the axis protein to form proteoglycans. In turn, the proteoglycans are associated with hyaluronan in the high volume supramolecular complexes (Figure 2.13). This structure could include more than hundred proteoglycan monomers, non-covalently attached to the single HA molecule. Such complex is stabilized by link proteins attached to the core protein of proteoglycan and hyaluronan. The supramolecular complex called aggrecan can have the molecular weight over 10^8 Da and can occupy a volume equal to the volume of a bacterial cell. Aggrecan is the main structural element of the cartilage matrix [41].

In a free state, the sulfated glycosaminoglycans are practically never found, but they do appear in pathologic conditions. As previously mentioned, chondroitin sulfate and sulfated glucosamine stimulate synthesis of proteoglycans and collagen in cartilage tissue, which influences proliferation of chondrocytes [66,67]. Similar to HA, many functional concentration-dependent features can be seen. With high amounts of free glycosaminoglycans in cultural media (150–200 mcg/ml), suppression of chondrocytes proliferation has been unambiguously found, but at the same time production of proteoglycans and collagen by chondrocytes increased [65]. In concentrations lower than 10 mcg/ml, the same product accelerates proliferation of chondroicytes, but does not influence the synthesis of collagen and proteoglycans. This is an excellent example of the competitive relationships of proliferation and differentiations: as cells start to divide, they stop synthesizing extracellular proteins.

Many functional properties of glycosaminoglycans manifest themselves in the ability of glycosaminoglycans to suppress the activity of enzymes that participate in the destruction of the connective tissue's cellular matrix. They suppress activity of hyaluronidase and granulocytes

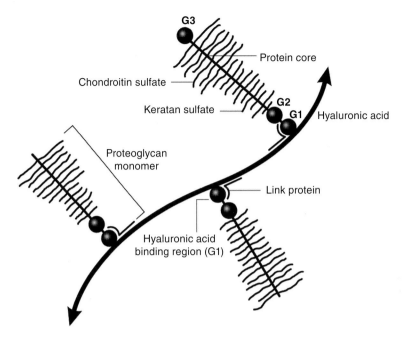

Figure 2.13 Participation of hyaluronan in formation of supramolecular complexes of proteoglycan aggregate. G1, G2 and G3 are globular folded regions of the central core protein [41]. Reproduced from [41]. International Labor Office, Geneva

elastases [69,70], collagenases [71], lysosome hydrolases and serine proteases [72] as well as other enzymes [59,74]. There is data about stimulation of wound healing by proteoglycans of cartilage [130,131] and other chondroitin sulfate decomposition products with a molecular weight of 4000 Da [132]. The decomposition products of collagen (peptides) activate collagen synthesis by fibroblasts as well. The action of peptides can be exerted through specific receptors for collagen molecules and peptides found in the fibroblast membranes [133].

The final products of decomposition can be used as the starting materials or precursors for synthesis of physiologically active substances. For example, sugars could be the precursors for synthesis of nucleotide sugars, amino acids for certain enzymes and so on. It is obvious that multiple products of biopolymer decomposition, due to the acquisition of new functional properties, try to 'regenerate' the initial state of cell tissue or form new stationary state. This is an example of the biological manifestation of the general principle, which is known in chemistry as Le Chatelier's principle. Obeying this principle of chemical and non-linear living systems (non-linearity means that there are several degrees of freedom in the development of the biological process) that possess multi-contour regulation (transfer system from one state into another) assures the transfer of the biological system from one stationary state into another one when the conditions have changed.

Under harsh conditions, when a cell cannot return to initial state or transfer into other stationary condition, a cell can enter into chain catalytic process of apoptosis. This is a controlled cell breakdown with a characteristic cascade activation of hydrolytic enzymes. The group of genes that codes the family of proteases (kaspases) plays a key role in the process of apoptosis through cascade activation of one enzyme by another. Similar properties are characteristic for the enzyme family of the hyaluronidases. Their concentration is regulated on the genetic level by the change in activity of the corresponding genes and at the metabolic level by glycosylation of the specific conservative fragments in different hyaluronidases [134]. To summarize these mentioned results, it is possible to suggest that a considerable part of cell decomposition products, such as fragments of polysaccharides, polypeptides and polynucleotides, acquire new physiological properties that allow for the biological systems to recuperate.

What are possible recuperation mechanisms? One could be activation or inhibition of the enzyme synthesis and activity. Low molecular weight fragments of biopolymers can likely create an 'informational field' for the regulatory systems by turning on specific receptors and/or forming a temporary extracellular matrix.

Decomposition of the protein-carbohydrate complexes in the extracellular matrix is catalysed mainly by lysosome hydrolases. Under normal conditions, the sulfated glycosaminoglycans are connected by covalent bonds with core proteins and are practically not presented in the free state. However, in pathologic conditions where the balance of synthesis-decomposition in the extracellular matrix is shifted to decomposition, free hyaluronan and sulfated glycosaminoglycans fragments appear and are then able to reveal additional unique properties. The fragments suppress activities of different enzymes that participate in decomposition of the extracellular matrix – hyaluronidases, elastases, collagenases, serine proteases, pepsin, lysosome hydrolases, acid cathepsins and other enzymes. They suppress cell apoptosis, neutralize action of free radicals and help to accelerate healing of wounds. For example, skin extracellular matrix hyaluronidase is able to hydrolyse dermatan sulfates in addition to hyaluronan. At a certain concentration of free fragments, dermatan sulfates suppress activity of hyaluronidase (Figure 2.12), thus forming a regulatory chain that controls the enzyme activity.

Intermediate products of the biopolymers decomposition are active participants of the cell signalling system. Obviously, this is characteristic for hyaluronan as well. Polypeptide fragments often possess hormonal activity. The most comprehensive description of hormonal protein decomposition, which results in a series of hormonal peptides, is described in [135]. At the decomposition of the nucleus protein (histone H4), an octapeptide with clear opioid properties is created. The monomer products of the biopolymers decomposition could be re-utilized into new polymers or into hormone synthesis.

Another poorly studied area is the use of intermediate biopolymer fragments in the construction of the transition structures. The intercellular matrix is a zone of self-assembling macromolecules which are exported from the cell. A connection has been declared between the extracellular matrix and the system of microfilaments and microtubes of the intracellular matrix through transmembrane proteins and their mutual influence. But since the extracellular matrix is involved in signal transduction, the signal paths between the extracellular matrix and cells plays a dominant role in the selection of cell adaptation direction according to environmental variations. The decomposition of HA by hyaluronidase of the extracellular matrix, as well as decomposition of sulfated glycosaminoglycans and proteins by the corresponding enzymes, likely leads to formation of the temporary matrix that is based on the intermediate fragments. This temporary matrix is capable of signal transfer. Also, its functioning can transfer a cell into another stationary state. During fractional biopolymer decomposition, it is possible that the formation of the multilevel three-dimensional structures of the extracellular matrix up to liquid-crystalline could occur. The suggested scheme of the intermediate temporary matrix formation is presented in Figure 2.14.

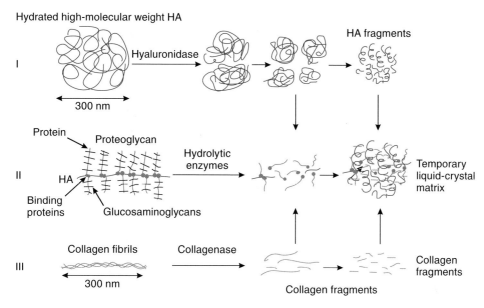

Hydrated high-molecular weight HA

HA fragments

I   Hyaluronidase

300 nm

Protein
Proteoglycan
Hydrolytic
enzymes

II   HA

Binding
proteins   Glucosaminoglycans

Temporary
liquid-crystal
matrix

Collagen fibrils   Collagenase

III

300 nm

Collagen
fragments

Collagen fragments

Figure 2.14   Subsequent milestones of intercellular matrix re-modelling during the decomposition of hyaluronan (I), proteolycans (II) and proteins (III)

The self-organization of HA into different three-dimensional structures in the presence and absence of the salts has been shown experimentally [128,136]. HA can form helixes of different structures, depending on its molecular weight, ionic solvent composition and water content (this will be discussed in detail in Chapter 4). It is suggested that even a double helix can exist under certain conformational conditions [137]. In aqueous solutions, HA adopts conformation with interchain hydrogen bonds. The light scattering characteristics show that the molecule behaves as a randomly banded and quite loosely packed chain with a bending radius of 200 nm. The level of chain packing and stiffness is associated with the presence of internal hydrogen bonds. Finally, the macromolecule of hyaluronan could be approximated as a highly hydrated sphere. In aqueous solution, the viscosity of the HA has maximum values as a result of its most extended conformation. When the concentration of HA in aqueous solution increases, the viscosity also rapidly increases due to the chain's weaving and formation of a three-dimensional network. This is how the gel-like structure is formed. The addition of the salts into aqueous solutions dramatically reduces their viscosity. Hyaluronan is a highly hydrophilic polymer. Each dimer contains a carboxyl group that dissociates at the physiological pH (7.0–7.4), thereby enhancing the polyanionic character of the polysaccharide even more. Polyanion can bind a considerable amount of metal cations such as sodium, potassium, calcium, magnesium and others because of its hydrate shell. As a result, the hyaluronan molecule can be increased in volume by 1000 times and form 1000 weakly packed hydrated matrixes. Understanding of the features of such physicochemical characteristics clarifies the physiological role of hyaluronan. Living systems use polysaccharide properties such as elasticity, viscosity, rheological properties (volume filler and joint lubrication), osmotic pressure (osmotic buffer is used to maintain tissue homeostasis), diffusion barrier (different diffusion rates of the differently size molecules), flow resistance (a thick chain network delays fluid flow) and excluded volume (a three-dimensional network of HA chains displaces other macromolecules).

It is believed that hyaluronan is capable of existing both in the extracellular matrix and on the cell surface in an incredibly large number of conformational states including elongated chains, helixes, 'slack' spirals, 'clips' and condensed rod-like structures. Interacting with each other, HA chains form fibrils, webs, piles and other shapes [47,123,137,138]. In the physiological solution the rigid helixes can be formed. Their size is random and could be on average about 2.5 mcm (perimeter length) on each chain with a molecular weight of 1 million Da. Such chains contain about 2650 disaccharide fragments. The secondary hydrogen bonds are formed along the polysaccharide axis. They assure stability and help to form hydrophobic domains. Due to these bonds, HA is organized into ordered structures.

The natural polysaccharides from different sources have molecular weights ranging from 5000 to 10 000 000 Da. The biopolymer in human synovial fluid, for example, has an average molecular weight of 3 140 000 Da. The HA macromolecule is energetically stable due to the stereochemistry of the constituent disaccharides. Bulky substituents of the pyranose rings are in the sterically favourable positions, while the hydrogen atoms occupy a less favourable axial position (see Chapter 4). Obviously, due to the fact that HA in the body is not subjected to additional chemical modifications after synthesis, the functional groups remain intact.

The variety of the conformational states of HA – their physical-chemical properties – depend on chain length, the surrounding ions and media pH, binding with other cell components to form a large spectrum of the physiological, functional and biological properties of the biopolymer. As was mentioned before, hyaluronan can reveal opposite biological functions depending on the size of the macromolecule. Oligosaccharides of low molecular weight hyaluronan can function as endogenous alarm signals and, depending on their molecular weight, can initiate different signalling pathways. The ability of low molecular weight fragments of HA to bind with the CD44 receptor on the plasmic cell membrane is an experimentally proven fact. Numerous data indicate a broad spectrum of specific HA functional activities that are dependent on size [47]. In particular, hyaluronan fragments of low molecular weight influence expression of the thermal shock protein [106]. Oligomers with 4–20 disaccharide units stimulate capillary growth and also participate in the control of cell homeostasis [126].

Testicular hyaluronidase hydrolyses the biopolymer to 4, 6, 8, 10, 12, 14, 16, 18, 20 and other disaccharide fragments [139]. Disaccharides are able to crystallize and other oligosaccharides can form liquid-crystalline structures. Thus, low molecular weight fragments of all biopolymers (hyaluronan, nucleic acid, proteins polyphosphate and other polycation and polyanion biopolymers) can form liquid-crystalline biological structures with intermediate states, acquiring new functional properties [140–143].

2.2.6 The Rules of Biopolymer Functional Cleavage

As a summary, we can formulate the following rules of the consecutive formation of the new functions during biopolymer decomposition:

Rule 1. Under controlled cleavage of the biopolymers, new oligomers with new functional properties are formed.

Rule 2. In the functional cleavage of the biopolymers, the cascade mechanisms of depolymerization are involved.

Rule 3. Upon biopolymer depolymerization, new products with antagonistic functions could be produced.

Rule 4. New functional properties of the intermediate cleavage products are aimed at the restoration of the biological system (Le Chatelier's principle in biology).

The important medical aspects of research, isolation and use of intermediate products of biopolymer degradation in medicines have emerged from these generalizations. In particular, products of hyaluronan degradation *in vivo* are considered bioactive natural products [139]. After cleavage from proteoglycans, free chondroitin sulfate prevents destruction of the extracellular matrix's collagen framework and supports the formation of a temporary matrix from partially degraded collagen. The creation of a temporary matrix with the aid of sulfated glycosaminoglycans serves to stop its further deterioration and ensures its more rapid substitution by the regular extracellular matrix [65].

Under severe exposure of gamma radiation, polypeptides and fragments of DNA circulated in the blood of Chernobyl liquidators for quite a long time. Based on the data from various sources, the concentration of hyaluronan fragments in the blood serum ranged from 0.01 to 0.10 mcg/ml. The half-life of endogenous HA is estimated to be 2.5–5.5 h. The radiolabelled HA, injected into the blood of animals, is destroyed quite quickly in the liver. The excretion of endogenous HA occurs through the kidneys. Decomposition products of the biopolymers in the tissues have been called cytoproteins, tissue hormones or biogenic stimulants by many authors, and are involved in the regulation of different cellular systems regionally. This type of autoregulation is based on the feedback between decomposition and synthesis of molecules and cells, both in physiological and pathological conditions [144]. Peptides are involved in regulation at all levels of body functioning, because multifunctionality is a characteristic feature of peptide bio-regulators [145,146]. It is assumed that short peptides can participate in regulation of apoptosis/necrosis [147]. Evidence accumulated to date suggests that this multifunctionality is inherent to hyaluronan oligosaccharides and other glycosaminoglycans that become bioactive natural products [106,137,139].

Thus, oligomeric fragments of natural polymers are involved in the regeneration of cells and tissues by activation of synthesis of new polymers, creation of temporary matrixes, delay of cell apoptosis and cytolysis and activation of cell proliferation. Many cells that do not grow in normal conditions start proliferation after injury to organs and tissues, after the splitting of different biopolymers and accumulation of their oligomeric fragments. All these mechanisms are designed to restore the biological systems, assist their adaptation to new conditions and form a new stationary state of the system.

2.3   Hyaluronan Signalling Systems

It has been determined that some cells begin to aggregate after addition of hyaluronan. The observation of aggregation was the first indication that the polysaccharide macromolecule can be bound with the cell surface. At the present time, several protein receptors, which specifically bind biopolymers on the surface of the cytoplasmic membrane, have been isolated. These have a high affinity to the HA receptors CD44 and the receptor for HA-mediated motility (RHAMM) [123]. Another receptor for hyaluronan endocytosis was found on the membrane of the endothelial cells of the liver sinus. It differs significantly from other proteins able to bind with hyaluronan. In order to understand their differences, we must examine the organization and functioning of the cell signal system and how they can, for example, transfer hormone signals [41,148].

The hormones act on the target cells by binding with appropriate receptors. By their chemical nature, the receptors are proteins and usually contain several domains. The concentration of hormones in the extracellular liquid is very low, in the range 10^{-6}–10^{-11} mmol/l. That is why it is important that the hormones and other signalling molecules have a high affinity to receptors. For the well-studied hormones, the constant of dissociation (K_D) of their complex receptor is usually in the range of 10^{-8} – 10^{-10} M and lower [149]. The localization of the receptors is different. The receptors of the peptide hormones and adrenalin are located in the cytoplasmic membrane, whereas receptors of the steroid and thyroid hormones are located in cytosol (glycocortioids) or in the cell nucleus (oestrogens, androgens and tireoid enzymes). The membrane receptors have three functionally different domains. The first domain is responsible for recognition and is located in the N-terminus part of the polypeptide chain on the outer part of the cell membrane. It is usually glycosylated and ensures hormone recognition and binding. The second is a trans-membrane domain that contains one or several alpha-helical fragments of the polypeptide chain. The third domain is located in the cytoplasm and creates a chemical signal inside the cell. This domain must associate the recognition of the bound hormone on the outer side of the membrane with the appropriate cell response, or in other words, change the quantity and/or activity of enzymes by its chemical modification (e.g. phosphorylation – dephosphorylation). Binding of hormone or other signalling molecules (primary mediators) alters the conformation of the receptor. The alteration is then transmitted to other macromolecules in the membrane or cytosol, leading to conjugation of one molecule with other in a process called signal transduction. As a result, a signal is generated, either directly or via secondary mediators (cyclic adenosine monophosphate, Ca^{2+}, NO, etc.). It regulates cell response by modifying the activity of enzymes or the amount of enzymes and other macromolecules. Thus, there are direct signalling pathways from the cell surface into the cytoplasm and nucleus. The activation or inhibition of gene activity, enzyme synthesis, cell proliferation, apoptosis, migration and differentiation of cells takes place through these regulatory circuits. Hyaluronan is involved in these processes by binding to the membrane receptor CD44. The structure of the receptor is shown in Figure 2.9.

When exogenic high molecular weight HA is added to very active transformed fibroblasts, the motor activity of the cells is sharply reduced by the polysaccharide binding to receptors CD44 [150]. If the fragment of the protein CD44, which is located on the surface of cytoplasmic membrane and specifically bound with HA, could be cleaved by the membrane-bound protease then the signal circuit would be broken and the migration activity of the cells would change [151]. Through the same receptor CD44, hyaluronan affects cell apoptosis and inflammatory response [152]. It is assumed that as a receptor of the endothelial cells and keratinocytes, CD44 can bind at least 6–10 oligosaccharide-long hyaluronan molecules [128]. Cell mobility is influenced through another HA receptor, namely *RHAMM*.

Another specific receptor for hyaluronan endocytosis has also been found, which has indicated that HA can be transported from the extracellular matrix into the cell by particular methods called 'receptor-mediated endocytosis' or 'absorptional endocytosis'. The macromolecules (ligands) are bound with specific receptors on the cell surface. Then the ligand-receptor complex creates a cluster in the area where an endocytosis vesicle (endosome) is formed to be further transferred inside the cell. This specific macromolecule transportation method is much faster than transportation due to liquid phase endocytosis. In addition, receptor-mediated endocytosis is a mechanism of the selective concentration of the signalling compounds, involving their capture and transfer inside of the cell. This is how several hormones, including insulin, polypeptides and cholesterol, are transferred inside the cell [153,154]. It was recently found that HA, which has a specific receptor on the surface of the cytoplasmic membrane, is transferred by the same mechanism. Receptor-mediated endocytosis usually leads to the transfer of intercellular molecules into lysosomes. It is possible that protein and polysaccharide signalling molecules, or the products of their cleavage, act further inside the cell similarly to steroid or thyroid hormones. Tetrasaccharides of hyaluronan probably act at the activation of the thermal shock proteins using the same mechanism [106].

In addition to structural and ion-exchange functions in the organization of extracellular matrix and its hydrodynamics, hyaluronan participates in signalling paths that regulate cell activation, proliferation, migration, adhesion, differentiation and apoptosis. To perform these fractions, HA cooperates with groups of proteins called hyalhedrins.

Hyalhedrins could be divided in three types: (1) soluble proteins, (2) proteins that bind hyaluronan with other polymers of extracellular matrix and (3) proteins, which act as cell membrane receptors for hyaluronan. Two receptors, hyalhedrins CD44 and RHAMM, are well known. The main receptor on the cell surface, that is, CD44, has very high affinity to biopolymers and is expressed by different cell types including fibroblasts, epithelial and some endothelial cells, keratinocytes and others. Figure 2.15 schematically shows HA's

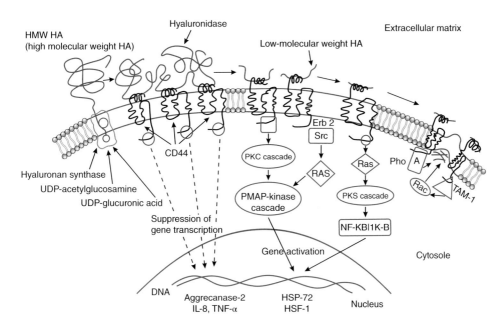

Figure 2.15   *Signalling scheme of transformation from hyaluronan through receptor CD44 and into the cell, nucleus and genetic system*

signalling path mediated by CD44 receptors. Binding of hyaluronan with CD44 can cause a cascade of the signals that affects the activation of gene transcription, switching off the 'differentiation' group of genes and turning on the 'proliferation' group of genes. This can change the cell cytoskeleton, leading to alteration of the migrational activity of cells. Cytoskeleton composition could be modified through the signal path with the participation of Rac- and RhoA-factors. The signalling cascade through protein kinases C and MAP-kinases can be involved with gene expression inside the nucleus. Protein kinases participate in the chemical modification of enzymes, namely in process of phosphorylation, or dephosphorylation of polypeptides. In response to phosphorylation-dephosphorylation, the enzyme is activated or deactivated [41]. In particular, hyaluronan synthases, being transmembrane proteins located on the cell membrane, contain cytoplasmic domains with sites that are targets for phosphorylation by protein kinases C and A. Stimulation of protein kinases A and C activity (e.g. phorbol esters activate protein kinase C) leads to activation of the synthesis of hyaluronan as well. Probably synthesis of polysaccharides of different molecular weights is regulated through the cascade mechanism, mediated through receptor CD44. The HA receptor is associated with different cell processes, including cell proliferation, motion and association, as well as HA degradation and re-utilization. The receptor RHAMM is localized on the surface of the cell membrane, as well as in cytosol and in the nuclei of different cells. Because of that, the receptor can participate in hyaluronan endocytosis. RHAMM is involved in the regulation of cell response to growth factors stimulation and plays a role in cell migration, especially of fibroblasts and smooth muscle cells.

Other pathways of the signal transduction initiated by hyaluronan are described in the literature. Medium size HA fragments with a maximum molecular weight of 200 000 Da, as well as the fragments composed of six oligosaccharides, activate mouse alveolar macrophages by mediating another path called NfkB/IkB [128]. At the same time, synthesis of mRNA, secretion of haemocrine proteins and induction of NO-synthase, are also enhanced. There is evidence that the transduction of the inflammation signals by the degradation products of hyaluronan is carried out through receptors TLR_2 or TLR_4, or through both receptors at macrophages and dendrite cells. It is believed that these toll-like receptors (TLR) belong to the innate immune system.

Most of the surface receptors, which are activated by extracellular signal ligands, produce intracellular signals by one of two basic methods: (1) they alter the activity of membrane-bound adenylate cyclase, which, in turn, changes the concentration of the intracellular mediator cyclic adenosine monophosphate (cAMP), or (2) they affect the permeability of ionic channels, which results in changing of the membrane potential and intracellular concentration of Ca^{2+} ions. A combination of these two methods is also possible. Cyclic adenosine monophosphate and calcium bivalent cation are called secondary intracellular mediators (messengers). Increased concentration of cAMP further activates intracellular protein kinases, which phosphorylates various enzymes in metabolic pathways, increasing or decreasing their activity [155–158]. These reactions have a short life because the specific enzymes (phosphodiesterases) quickly destroy any excess of cAMP.

Another secondary messenger of signalling systems is Ca^{2+} cation. The opening of calcium channels in some secretory cells releases calcium ions from internal stores (mitochondria, endoplasmic tanks), and may trigger exocytosis. Calcium channels are also open only for a short time and calcium ions are quickly pumped out and/or bind with intracellular molecules, such as calmodulin, changing their activity [159,160]. Hyaluronan and its oligosaccharides trigger a cascade of signals through these secondary mediators.

It should be noted that the results of various studies, which are the basis for the presentation of the signalling pathway schemes, are complex and quite contingent. They represent only simplified, suggested ways and mechanisms of complex process regulation. Nevertheless, such schemes are important for systematization and have heuristic value.

In biological processes, the significance of hyaluronan is not limited to its ability in physical interaction, but also for its ability to participate in regulation systems that can trigger a signal transduction from the extracellular matrix to the nucleus and the cell genetic system. Thus, HA is involved in the subtle regulation of cellular activity. Signalling pathways today are not considered as linear sequences of biochemical modifications, but rather as networks of the varying levels of complexity. These networks are initiated by extracellular ligands, which interact with other receptors of the cell surface. The complexity of the process is due to organization of these receptors. Irregularity of the receptor expression or dysfunction, including CD44, leads to numerous pathologies.

The general scheme of transmitting signals represents the interactions between intercellular ligands and receptors that affect the interaction between cell surface proteins and further the interaction of cytoplasmic proteins, which, in turn, regulate gene transcription (Figure 2.15). The uniqueness of hyaluronan receptor CD44 is that it can bypass the last stage of the path and activate a gene transcription by itself. This is done by means of proteolytic processing of the receptor CD44 in the following way: First, the membrane metalloprotease can cleave the domain of the protein CD44, which is on the outer surface of the cytoplasmic membrane and serves as a hyaluronan binding site [151]. Because of the ectodomain cleavage, cell migration is facilitated. In fact, this is a mechanism of cell migration regulation by CD44. Further cleavage from the receptor protein CD44 by the proteases leads to domain transportation into the nucleus as well as its combination with other transcription factors (particularly with *c-Fos* and *c-Jun*). It also enhances transcription of several genes and induces the expression of the gene responsible for the synthesis of protein CD44. It is assumed that this regulatory circuit contributes to a more rapid turnaround of CD44 protein required for the control of cell migration [74]. When membrane-bound metalloproteinase inhibitors block this regulatory circuit, consequently the acceleration of the receptor CD44 turnaround is not observed.

The receptor CD44 is presented on the cell surface of most vertebrates. These are a polymorphic group (family) of proteins with a molecular weight of 80 000–200 000 Da. The proteins of CD44 are encoded by a single gene. Gene CD44 is highly conservative; the protein family is produced by the mechanism of alternative splicing, which concerns mainly the extracellular, stem structure of the protein. The first five unchanged exons encode the N-terminal globular ectodomain, which contains a fragment of 90 amino acid residues (from 32 to 123). These fragments are used for the binding of HA and other glycosaminoglycans. CD44's affinity towards glycosaminoglycans depends on posttranslational modifications of the protein ectodomain (e.g. glycosylation). Such modifications are dependent on the cell types and growth factors. The CD44 receptor binds HA if the polysaccharide consists of more than 20 residues. Binding occurs with more than one CD44 molecule [74].

N-terminal globular ectodomain is separated from the plasma membrane by a short (46 amino acids) stem structure that contains a putative proteolytic cleavage site (Figure 2.9, 2.15). Tumour cells often express CD44 varieties with large stem structures that undergo O-glycosylation. A transmembrane region is composed of 23 hydrophobic amino acids and cysteine residues. A cytoplasma tail is often connected to the proteins and participates in ensuring CD44's association with the cells' skeleton, particularly with the protein ankyrin that provides contact with the cytoskeletal protein spectrin. These interactions are involved in providing HA-dependent cell motility and adhesion.

The binding of ankyrin with CD44 is controlled by phosphorylation of serine residues of Rho-kinase. The absence of this active enzyme removes the possibility of CD44 phosphorylation, membrane fluidity, cell migration, and metastasis of tumour cells. The proteins ezrin, radixin and moesin interact with the basic amino acid motif of the cytoplasmic tail and cross-link the actinic cytoskeleton to the CD44. These proteins are important for the regulation of cell shape and migration, and their binding is regulated by the phosphorylation-dephosphorylation of serine residues in the cytoplasmic tail of CD44 by protein kinase C. Various amino acid reactions of phosphorylation-dephosphorylation (serine, tyrosine and threonine) in the C-terminus's cytoplasmic tail by various protein kinases can affect both the intramolecular interactions between the N-and C- protein terminus in CD44 and the protein-protein interactions. For example, the dephosphorylation of several proteins is initiated either at high cell density or by the addition of high molecular weight hyaluronan [74].

The precise molecular mechanisms by which CD44 performs its functions remain mere speculations. Whether it acts as a ligand (binding different component receptors of the intercellular matrix), as a specific platform for growth factors and matrix metalloproteases, as a co-receptor for modulating of the growth factors receptors, or as an organizer of the cortical actin cytoskeleton (Figure 2.15).

The protein CD44 can be involved in the formation of receptor clusters. While CD44 interacts with various components of the extracellular matrix, such interaction is not essential, except for its interaction with hyaluronan. CD44 affinity to hyaluronan is likely modulated by extracellular factors. In particular, it increases by mitogenic stimuli and is dependent on the level of glycosylation of the extracellular domain. It is also possibly dependent on phosphorylation of specific serine residues in the receptor cytoplasmic tail. Modulation of the binding affinity of CD44 to the polysaccharide is important for the cells' migration in the hyaluronan-rich extracellular matrix. The evidence is that CD44 is located on the outer edge and the release of the cell that had been bound with hyaluronan is prompted by lamellipodia of some types of migrating cells and cleavage of the CD44 ectodomain by proteases. The suppression of protease activity, which cleaves the CD44 ectodomain, leads to inhibition of tumour cell migration. The binding of HA with CD44 initiates the metabolism of hyaluronan. CD44 is involved in some 'passive' functions that do not require direct activation of a signalling cascade. In fact it is known that HA can act as an adhesive molecule, creating the bridges between the cells that express CD44.

The proteins of CD44 can 'catch' and concentrate growth factors as well as connect the substrates and enzymes (i.e. behave as platforms for organization of the different processes) where long-distance and short-distance signalling events could be concentrated on the cell surface.

Many extracellular stimuli are perceived by a cell through the receptor tyrosine kinases. Their cytoplasmic domains have kinase activity and phosphorylation motifs, which, after receptor activation, create sites for multiple components of the intracellular signalling network. Receptor activation is achieved by mutual intermolecular autophosphorylation of receptor subunits. This association is induced or stabilized in the receptor by ligand binding.

Activation is more complex for receptors that lack the kinase domain. To activate these receptors, they cooperate with the receptor tyrosine kinases and are called co-receptors. In CD44 there is no catalytic activity at the cytoplasmic end. It functions as a co-receptor with respect to the family of receptors, which have tyrosine kinase activity. Many intracellular proteins, which are involved in signalling, are associated with the cytoplasmic tail of the CD44 receptor (Figure 2.9). However, the functional role of these associations is unclear. The only data that points to a direct signal transfer through the CD44 is presented in a study wherein treatment of cells with mitogens causes proteolytic cleavage of the CD44 intracellular domain. It is found in the nucleus and modulates the activity of certain genes in a complex with other proteins [74]. CD44's ability to interact with proteins, which are connected to the actin cytoskeleton, points to another mode of transmission of cellular signals from the extracellular HA into cell through CD44.

It was found that overexpression of CD44 cytoplasmic tails leads to a strong suppression of certain signal transfers. Signalling requires links of CD44 with actin and local cytoskeleton organization. The process of signalling is structurally organized; preformed components of signal transduction modules can operate in association with the actin cytoskeleton, which is connected through a series of proteins and receptor CD44.

Summarizing the numerous works on the structure and functioning of the protein family CD44, there are three main avenues of function:

1. CD44 could bind extracellular matrix ligands (primarily hyaluronan), which has an effect on cell behaviour regardless of their interaction with receptor tyrosine kinases, and the actin cytoskeletons can act as platforms for growth factors to participate in the function of the extracellular matrix (based on hyaluronan), in its assembly and dismantling.
2. CD44 may function as co-receptors that mediate signal transduction from the receptor tyrosine kinase.
3. CD44 may provide a link between the plasma membrane and cell cytoskeleton.

Cell surface proteoglycans such as syndecans and glypicans behave as co-receptors. Thus, a fibroblast growth factor binding to heparan sulfate chains promotes growth factor interaction with its receptor. Conversely, fibroblast beta-growth factor interaction with decorin suppresses its activity. The binding of growth factors and cytokines with the proteoglycans of the extracellular matrix suggests that these macromolecules can be concentrated in certain areas of the extracellular matrix and then released locally when needed. Thus, the extracellular matrix may serve as a depot of regulatory molecules, and hyaluronan and its fragments provide an informational system that signals on the state of the extracellular matrix [161].

2.4   Hyaluronan Functions in the Extracellular Matrix

Hyaluronan performs its major functions in extracellular matrix [41,162–164] where it fills the extracellular matrix upon its structuring through interactions with proteins [57,165]. Hyaluronan plays an important role in cell and tissue hydrodynamics, ion exchange and transportation of the different compounds in two opposite directions (from the blood vessels into the cell and from the cell into lymphatic vessels). It also participates in the active exchange of metabolites and ions and gases (O_2 и CO_2) between blood and tissues, and performs the role of 'structural mediator' during the cell interaction, creating channels for their migration. There is approximately 15 g of hyaluronan in the body of an average adult person [166], mainly in skin (dermis and epidermis), the synovial liquid of joints, the eyeball and other connective tissues. The turnover rate of HA is quite remarkable. During one day, approximately 5–7 g of hyaluronan is cleaved (hydrolysed) in the human body, which is half of the total amount. The same amount is synthesized [167].

In order to discuss the mechanisms of HA functioning in the extracellular matrix of different tissues, we will make a short excursion into the biochemistry of the extracellular matrix.

2.4.1 Extracellular Space

The cytoplasmic membranes of adjacent cells in various tissues are associated with each other by cell-type connections of the tight junctions, gap junctions and desmosomes. Most cells are surrounded by extracellular matrix, a complex network of interconnected macromolecules among which hyaluronan has an important place. It is possible to speculate about the size of intercellular (interstitial) space of a human by the distribution of the liquid in the two major intracellular and intercellular compartments. The total water content in the body of an adult weighing 70 kg is approximately 42 l. Without blood plasma, various sources estimate the intracellular fluid volume to be approximately 28–30 l, and intercellular fluid volume to be 10 l [168,169]. The most detailed study was performed for brain extracellular space, which takes 12–14% of the total organ volume. The neurons and glial cells are separated by intercellular gaps which have widths of 15–20 nm. The slots are interconnected, forming extracellular space filled with a liquid through which the exchange of substances between the neural and glial cells by diffusion takes place [168]. The extracellular space helps for sodium-potassium exchange between the neurons and extracellular media and, consequently, conduction of nerve impulses. During embryonic brain development, the extracellular matrix contains a large amount of hyaluronan. However, in time, the content in the extracellular matrix of the brain dramatically decreases. In addition to water, extracellular space contains a significant amount of biopolymers, molecules, and ions which are united in the concept of extracellular matrix.

2.4.2 Composition and Functioning of the Extracellular Matrix

Extracellular matrix (intercellular, pericellular matrix) is composed of a variety of polysaccharides and proteins that are spontaneously organized into ordered structures. It represents a supramolecular complex, formed by a network of molecules connected to each other [31,41]. In the human body, the extracellular matrix produces highly specialized structures such as cartilage of the tendon, the basal membrane and after the secondary deposition of calcium phosphate, bones and teeth. Since these structures differ both in molecular composition and methods for organizing the main components, they comprise the various forms of the extracellular matrix [31,41]. Here we draw a generalized 'picture' of the extracellular matrix and the functional role of hyaluronan in it.

The extracellular matrix is an area of biopolymer self-assembly and these are exported to the extracellular space by fibroblasts or the cells of this family (chondroblasts in cartilage and osteoblasts in bone tissue). The molecular structure of the extracellular matrix is influenced by macrophages (histiocytes), basophils (mast cells or labrocytes), mesenchymal cells, adipocytes (fat cells), pericytes and transient cells, which migrate into the connective tissue in response to specific stimuli – lymphocytes, neutrophils, basophils, eosinophils and plasma cells [133].

The extracellular matrix comprises mainly fibrous protein structures of collagen and elastin, embedded into a hydrated polysaccharide gel of HA and other glycosaminoglycans [41,162–164]. The macromolecules of hyaluronan, other glycosaminoglycans and proteoglycans form a highly hydrated 'basic substance' gel of the extracellular matrix, in which the collagen fibres are plunged [64]. The collagen fibres strengthen and streamline the matrix, and the aqueous phase of the polysaccharide gel ensures the diffusion of metabolites, hormones, signalling molecules and gases between the blood and tissue cells. Intercellular water could be estimated to comprise 25% of the entire water of the body, out of which about 5% is in the circulatory system and the rest in the intercellular matrix. The intracellular water accounts for about 40%.

A constant exchange of liquid substances, ions, gases can be seen between these three volumes. In the body, the main parameters of liquid, such as pH, osmotic pressure and volume, are strictly controlled. The pH levels of the extracellular fluid and blood are the same, and are normally about 7.36. Several biological conformations of macromolecules and the catalytic activity of enzymes are dependent upon pH. The osmotic pressure of the extracellular fluid and blood plasma are similarly dependent. Hyaluronic acid of the extracellular matrix plays a very important role in maintaining these parameters within the physiological norms. As an intermediate between the external media and the cells, extracellular matrix performs the coordinative functions between the cells and different cellular structures. The functioning of the cell cytoplasmic membrane is directly dependent on the extracellular matrix.

Differences in the chemical composition of the membrane external surface (the extracellular matrix is an extension of it) and the internal surface creates membrane structural asymmetry. Membrane surfaces that face the cytoplasm and the extracellular matrix are hydrophilic and those at the central part of the membrane are hydrophobic. Carbohydrates are usually attached to proteins and lipids, but in any case they are located on the outer rather than the inner membrane surface. Hyaluronan as well as other membrane proteins contribute into membrane asymmetry. In membrane, the protein/lipid ratio in membranes varies in the range from 1:4 to 4:1, but the most typical ratio is 1:1. Protein molecules embedded in the membrane or attached to them are oriented in a certain way. Some proteins are localized on the inner membrane surface and the ratio of their hydrophilic and hydrophobic amino acids is similar to that of cytoplasmic proteins. Usually secreted proteins undergo glycosylation, whereas cytosolic proteins generally do not. Proteins on the outer surface of the membrane are closely associated with the carbohydrate chain of the extracellular matrix, HA and sulfated glycosaminoglycans. Integral proteins are embedded into the double lipid layer of the membrane by a hydrophobic core and have open areas (sites) at both sides of the membrane. The fragment of the integral protein, which enters the intercellular matrix, is composed of hydrophilic amino acids and is usually glycosylated. The cytoplasmic site of an integral protein is also hydrophilic and either performs enzymatic functions or binds the enzymatic protein. The matrix and cytoplasmic fragments of integral proteins excludes the possibility of membrane protein inversions and slows the lateral diffusion of the membrane lipid molecules. The lipid molecules in the membrane may move freely in their own layer in the side (lateral) direction by changing position up to a million times per second. However, the transitions in the third dimension (from one bi-lipid layer to another one in a flip-flop movement) are strictly limited and occur no more than once a month. A lipid may pass the distance of 1/40 of the average diameter of the cells in its layer in 2 s, but the probability of passing to another layer is extremely small. Such an event could happen no more often than once every 100 s [170]. Those occasional passing results in the discovery of an important property of the membranes: their components can be rearranged and clustered to form signalling systems, receptor associations or facilitate mutual transport, without mixing internal and external interfaces. Existence of two main phases of the membrane is assured: a gel phase and a liquid crystal phase [171]. At the same time there are possibilities for sol-gel transitions in the cell cytosol.

Fragments with different phases can coexist in one membrane. The stable and highly ordered fragments of the membrane (rafts) serve for the formation of signalling cascades, such as the stimulation of proliferation. Liquid crystalline zones participate in exo- and endocytosis [172]. They create structural asymmetry of the cytoplasmic membrane and provide its functional asymmetry, which means that the inner and outer surfaces of the membrane must function differently. Otherwise, molecules or ions that are transferred from the intercellular matrix into the cell could be transported back to the extracellular matrix elsewhere. Asymmetry can also be observed in other properties of the membranes. Thus, hormone receptors and other signalling molecules are associated with the membrane and serve for transmission of signals from the extracellular matrix into the cells [173]. Binding of receptor with a signalling molecule alters the conformation of the transmembrane receptor protein that activates or inhibits the enzymatic activity of its cytoplasmic domain. This is the realization of the receptor-conformational regulation principle. This primary principle is the basis of all forms of intercellular regulation. More sophisticated signalling systems are built on a variety of proteins that transmit signals according to the cascade principle. In all cases, whether it is an enzyme, receptor or regulatory protein, the protein molecule can recognize a specific factor and interact with it accordingly, changing its configuration. In multi-component complexes, conformational changes of receptor molecules are cooperatively transmitted to the entire complex, affecting its functional activity. In the course of evolution, the intracellular regulatory systems emerge first and govern regulation levels such as enzymes, membranes and genetics. With the arrival of multicellular organisms, the intercellular, tissue and organ systems of regulation evolve and improve. They include hormonal, trophic and electrophysiological systems. Through the regulatory system, hyaluronan is involved in ensuring homeostasis in cells, tissues and organs [126] (i.e. in maintaining the constant parameters of the internal media of the organism) and also creates the conditions for its development (epigenesis). At all levels of the organization, homeostasis is assured by negative feedback, but epigenesis (embryogenesis and regeneration) is predominantly assured by positive feedback. All systems of regulation are closely linked and organized into a single system of regulation based on the principle hierarchy. Hyaluronan is built into the system and can induce both positive and negative feedback. Thus, the polysaccharide with a molecular weight of more than 500 000 Da slows cell proliferation, but low molecular weight HA (50 000–100 000 Da) activates cell proliferation. On the surface of the cytoplasmic membrane the hyaluronan macromolecule can be bound with specific proteins receptors CD44 and RHAMM [123] through which it initiates a cascade of intracellular signals. Proteoglycans often behave as receptor proteins as well. They bind to cytokines and other signalling molecules and regulate metabolism, gene activity and cell proliferation. For example, the fibroblast growth factor FGF-beta can stimulate the synthesis of hyaluronan in cell culture via receptors in two ways: (1) by activating the genes of hyaluronan synthase; (2) by activating hyaluronan synthases by phosphorylation through protein kinase C activation [150,174]. Based on several indicating features of specialized signalling molecules such as action longevity, lifetime (half-life), the affinity for the receptor and others, hyaluronan can be attributed to both the cellular (which deliver the signal from intercellular matrix into the cell) and intracellular signalling molecules. The signals from the extracellular matrix into the cells determine the intensity of metabolic processes, procedures and activities for an extraction of genetic information, the activity of matrix biosynthesis of RNA and proteins [106] and the sequence and rate of transport of macromolecules from cells into the extracellular matrix [175]. Thus, in the relations between the cell and the intercellular matrix regulatory chains with participation of HA are formed, providing coherence and mutual process control of the synthesis-decomposition and their dynamic stationary condition. With the appearance of hyaluronan in the extracellular matrix, the degree of asymmetry of the external and internal cell space is increased. Asymmetry of the cytoplasmic cell membrane is assured by differences in the chemical composition of the molecules on external and internal membrane surfaces, such as lipids, proteins and mainly polysaccharides, which are associated with proteins and lipids on the external surface of the membrane. Embedding HA, other soluble polysaccharides and glycosaminoglycans into the intercellular matrix increases the asymmetry degree of the membrane and both intra- and intercellular cell compartments. The intracellular compartment contains significantly less of the soluble polysaccharides compared to the intercellular. The degree of membrane asymmetry varies in different types of membranes and can be different during different stages of cell life [31,41]. Asymmetry is a prerequisite for proper functioning of membranes, for example, the unidirectional flow of different compounds only from a cell or out of cells. Further, HA of sufficiently large linear dimensions (up to 2.5 µm) carries out the functions of intercellular communications between the cells separated by extracellular matrix. The ability of HA molecules to bind with 1000 times more water than they weigh and at the same time to create hydrodynamic volume 10 times larger than the space occupied by the non-hydrated molecules, facilitates migration, adhesion, sorting of the cells during embryogenesis and damage regeneration in the adult organism that repeats certain steps of embryogenesis. For example, in case of bone damage (fracture), the cartilage tissue is initially formed and then replaced by bone tissue [45]. All of these functional properties of hyaluronan contributed to the emergence of a new branch of chordates in the phylogenesis and even today contribute to the preservation and evolution of the most advanced animal species.

2.4.3   The Role of Hyaluronan in Transportation of Substances through the Extracellular Matrix: Diffusion, Osmosis, Electro-Osmosis and Vesicular Transportation

Large blood vessels (arteries, veins) pass through subcutaneous adipose tissue. Derma has quite a developed capillary network, but epidermis has neither blood nor venous vessels. The same situation could be found in cartilage tissues. Therefore, transportation of substances from blood capillaries into the epidermis, cartilage and other cell tissues passes through the intercellular matrix. The only difference is the distance from the capillary to the cell and structural features of the matrices in different tissues.

During the transportation of gases from blood capillaries, oxygen passes into the intercellular matrix and penetrates the membrane, cell cytosol, and lastly the mitochondria double membrane, where it finds the final target for oxidation. Such a transportation mechanism could be called 'passive diffusion' as it assures the molecule transfer from the high concentration area, that is, due to pressure gradient (Fick's law). In 1 h, 190–750 ml of oxygen can be transported, depending on temperature and physical activity. Since the extracellular matrix is saturated with water, the transportation rate of the gases depends on their solubility in liquid media (Henry's law). Based on the solubility, the diffusion coefficient of carbon dioxide is higher than that of oxygen by 2.7 times. The amount of water in extracellular matrix directly depends on hyaluronan content as well as other glycosaminoglycans, which influences the transportation rate of gases. Following all gas laws, the colloidal system of the extracellular matrix influences gas exchange. However, this data is scarce and contradictory.

Most water in the extracellular matrix is bound and could be seen as a component of hyaluronan structure, sulfated glycosaminoglycans, proteoglycans and glycoproteins, which appear as macromolecular polyanions that form a web structure. In the physiological conditions the negative charges of such polyanions could be neutralized by the cations of sodium, potassium, calcium and other osmotically active ions that are surrounded with hydrate shells. As a result, gel-like and highly hydrated structures are formed. In these structures, water and ions undergo a quick exchange with formation of a stationary state. Similarly, water is bound in the intracellular space (cytosol). The membrane, which divides intracellular and intercellular spaces, is an active regulator of electrolyte transportation and creates an ion concentration gradient between these two compartments. Sodium is the main cation of the intercellular matrix (about 144 mM), and concentration of the ions of potassium, calcium and magnesium do not exceed 1–1.5 mM. Chlorine is the main anion of the extracellular matrix (about 114 mM). The main cation inside the cell is potassium (about 160 mM), then cations of magnesium (13 mM) and sodium (10 mM). The anions of the intracellular liquid are represented by proteins (8 mM), phosphates (50 mM), sulfates (10 mM) and bicarbonates (11 mM) [168,169].

Hyaluronan and glucosamine of the extracellular matrix possess functions such as water depositing and regulation of the diffusion rates of compounds in and out of the cells. They can also work as ion exchange media. The macromolecule transport through the hydrated web matrix is quite difficult. It is evident that the naturally 'invented' vesicular transport is based on the exocytosis mechanism [176] for biopolymer delivery (from the cell into extracellular matrix) as well as the endocytosis mechanism (from the extracellular matrix into the cell). Sulfated glycosaminoglycans are transported into the intercellular matrix via exocytosis mechanism, not as free molecules but as proteoglycan aggregates. Hyaluronan is not transported via exocytosis procedure, but is rather 'pushed out' of the cell through the cytoplasmic membrane by the enzyme that synthesizes it. In order to penetrate into the cell, hyaluronan uses the endocytosis mechanism. A possible mechanism of hyaluronan endocytosis is that it binds with the several receptors on the cell wall. During depolymerization, hyaluronan fragments can adopt different conformations and the receptors bound with hyaluronan approach each other (clustering). The receptor aggregation initiates the formation of an endocytosis vesicle that transfers the receptors and bound hyaluronan fragments into cell cytosol [45,176]. Using this mechanism, hyaluronan and its fragments can simultaneously regulate the transportation of different substances into a cell through the endocytosis path.

2.4.4   Hyaluronan in the Extracellular Matrix of Different Connective Tissues

A supramolecular complex of the intercellular matrix is formed by the complex network of the macromolecules that are connected to each other. In the different forms of the intercellular matrix, these structures differ from each other by both their molecular composition and organization of the main components, which are proteins and polysaccharides including HA. In the extracellular matrix of skin, the majority of hyaluronan is located between the fibres of collagen and elastin [177,178]. There is 0.1 mg of HA in 1 g of raw tissue of epidermis, but 0.5 mg/g in derma. Hyaluronan is connected with hyaladherins (proteins that have sites of specific binding with polysaccharides) and forms a polymeric web that fills the extracellular matrix. Hyaluronan, saturated with water, forms the gel structures in the extracellular matrix, and therefore maintains a high level of skin elasticity and acts as a selective filter that regulates the diffusion rate of the compounds that differ by molecular weight, hydrodynamic volume and charge. In addition, hyaluronan has some protective properties in skin, preventing penetration of microorganisms through the wounded surface. The skin's aqueous balance is maintained by two opposite-directions of flow: the diffusion of water into the dermis from the blood capillaries and its evaporation through the stratum corneum (horny layer). Keratinocytes of the skin epidermis are able to synthesize HA and their growth factor stimulates the synthesis of hyaluronan in keratinocytes by activation of hyaluronan synthase-1 and -3 [179]. Diffusion and evaporation are passive processes. Therefore, dermal hyaluronan has a particular importance in water exchange. Using its ability to bind large amount of water, HA plays a major role in maintaining the hydrated state of the dermis and epidermis. The horny layer of the epidermis is the barrier for water evaporation and it is also able to draw moisture from the air and hold it due to HA and other hygroscopic structures. Because of dehydration, skin loses elasticity and develops wrinkles. For the first land animals, moisture retention in the skin was equivalent to survival. Therefore, in evolutionary terms, HA apparently contributed not only to the appearance of the first branch of aquatic chordates but also their transit onto the land. In the cartilage tissue, the extracellular matrix occupies quite a large volume. It has the main functions of this connective tissue, including the spring function, a result of its elasticity and resilience properties. The main components of extracellular matrix, such as strong collagens, elastins, adhesive proteins and 'the main compounds' – hyaluronan and proteoglycans – could be found in it. One feature of the intercellular matrix is that contrary to dermal fibroblasts, the cartilage chondrocytes synthesize significantly less HA. In addition, in the cartilage matrix, hyaluronan forms large supramolecular complexes.

In these complexes the polysaccharide occupies the central enlarged part of the complex where the molecules of aggrecan (proteoglycan) are located perpendicularly to the axis with the interval of approximately 10 monosaccharide units (Figure 2.13). Aggrecan is the main proteoglycan of the cartilage matrix and comprises 10% by weight of the tissue and 25% by dry weight of the cartilage matrix. This proteoglycan monomer has a molecular weight of about 2 500 000 Da. The molecule is composed of carbohydrates (93%) and proteins (7%). The peptide chain is located in the centre of the monomer and is called an *accepting protein*. It consists of the amino acids residues of serine, glycine, glutamine and alanine. The serine residues are covalently attached to up to 100 chains of chondroitin sulfates and about 30–60 chains of keratan sulfates through trisaccharide xylose-galactose-galactose. Therefore, the aggrecan molecule resembles the shape of a bottle brush. Chondroitin sulfate consists of 20–70 disaccharide units and has a molecular weight of 5000–50 000 Da, and keratan sulfate includes 25 disaccharide units and has a total molecular mass of 4000–19 000 Da. In the cartilage matrix, the molecules of aggrecan are aggregated with hyaluronan by a small link protein (Figure 2.13). Both proteins connect to HA by non-covalent bonds in the C_1 domain. This domain interacts with approximately five disaccharide units of the polysaccharide. Then the complex is stabilized by the link protein. The C_1 domain and the link protein occupy 25 disaccharide units of hyaluronan. The final supramolecular complex contains one, approximately 1200 nm long molecule of hyaluronan, to which approximately 100 aggrecan molecules with the length of 300–400 nm and the same amount of the link protein are attached. The distance between two aggrecan molecules attached to HA is 25–50 nm. Aggrecan and link proteins are produced by chondrocytes. These components could interact with each other inside the cell, but the assembly process is completed in the extracellular matrix. It takes approximately 24 h to build a functionally active triple complex [41,170,177].

Small proteoglycans are included in the cartilage matrix as well. The matrix contains a small core protein to which one or two glycosaminoglycans chains are connected. Such proteoglycans include decorin, biglycan, perlecan, lumican and fibromodulin. Small proteoglycans are multifunctional macromolecules connected with other components of extracellular matrix and are able to affect their structure and functions. For example, decorin and fibromodulin can attach to fibriles of the type II collagen to limit their diameter. Decorin and biglycan connect with fibronectin and suppress their cell adhesion, and also connect with the tumour growth factor beta to reduce its mitogenic activity. There is data that confirms small proteoglycans have a regulatory role in the process of connective tissue regeneration. The main structural components of the intercellular cartilage matrix are type II collagen, aggrecan, hyaluronan (1 mg/g of the raw tissue) and water. Along with the small proteoglycan, there are types VI, IX and XI collagens, binding proteins, cartilage oligomeric proteins, chondroadherin, ankyrin and growth factors as well.

The fibrillar network of cartilage matrix is composed of types II, IX and XI collagen, providing cartilage strength. The collagen fibrils are connected with HA, which, in turn, forms supramolecular complexes with aggrecans. The high molecular weight aggregates composed of hyaluronan and aggrecan can be considered as polyanions due to a large amount of carboxyl groups, which provide a high level of hydration of the cartilage matrix and ability to perform spring functions.

The water content in joint cartilage is not permanent [180]. Under pressure, water leaves cartilage and when pressure is removed, water returns. In the morning, the water content of the intervertebral discs of cartilage is approximately 75% of the disk mass. Under pressure during the day, water content becomes approximately 20% less. That is why a human is taller in the morning than in the evening. The difference is approximately 1–2 cm. Also, due to these reasons, in zero gravity conditions it was found that astronaut height increases by up to 5 cm.

The extracellular matrix of bone tissue is characterized by a high level of mineralization. It includes 50% of the inorganic compounds by mass, mainly in form of hydroxyapatite, 25% of the organic compounds, and 25% of water. The main protein of the bone matrix is type 1 collagen (90–95%). The organic compounds are represented by a small amount of proteoglycans, including a carbon part that includes dermatan sulfate and keratan sulfate. Proteoglycans, together with other proteins (osteocalcin and sialoglycoproteins), are calcium-binding proteins and, therefore, participate in creation of a deposit of calcium and inorganic phosphate. It is possible they also participate in the control of the every-second modulation of cell calcium levels, as well as calcium and phosphorus levels in the blood serum. In the embryonic period, skeleton cartilage 'models' are gradually replaced by bone tissue. When the bones are injured and/or fractured, the regeneration begins with the replacement of damaged cartilage tissue followed by bone tissue replacement. The role of hyaluronan in cartilage matrix was discussed earlier. The functions of HA in the replacement of cartilage tissue with bone tissue itself is unclear. Based on the intensive studies of the structural, metabolic and regulatory properties of hyaluronan, at the present time the main medicinal applications of hyaluronan have been shaped (see Chapter 6).

References

[1] Shnol, S.E. (1979) *Physical and Chemical Factors of Biological Evolution* (in Russian). Nauka, Moscow. Шноль С.Э. (1979) *Физико-химические факторы биологической эволюции*. Наука, Москва.

[2] Anisimov, A.A. (ed.) (1986) *Fundamentals of Biochemistry* (in Russian). Vysshaya Shkola, Moscow. Анисимов, А.А. (ред) (1986) *Основы биохимии*. Высшая школа, Москва.

[3] Yarygin, V.N. (ed.) (2000) *Biology* (in Russian). Vysshaya Shkola, Moscow. Ярыгин, В.Н. (ред) (2000) *Биология*. Высшая школа, Москва.

[4] Oparin, A.I. (1924) *The Origin of Life* (in Russian). Moskovski Rabochii Worker, Moscow. Опарин, А.И. (1924) *Происхождение жизни*. Московский рабочий, Москва.

[5] Oparin, A.I. (1959) *Life, its Nature, Origin and Development* (In Russian) Nauka, Moscow. Опарин А.И. (1959) *Жизнь, ее природа, происхождение и развитие*. Наука, Москва.

[6] Oparin, A.I. (1966) *The Origin and Initial Development of Life*. (In Russian). Medicina, Moscow. Опарин, А.И. (1966) *Возникновение и начальное развитие жизни*. Медицина, Москва.

[7] Oparin, A.I. (1952) *The Origin of Life*. Dover, New York.

[8] Oparin, A.I. (1957) *The Origin of Life on Earth*. Academic Press. New York.

[9] Oparin, A.I. (1964) *The Chemical Origin of Life*. Charles C. Thomas Publisher, Springfield.

[10] Oparin, A.I. (1965) *The Origins of Prebiological Systems*. Academic Press. New York.

[11] Oparin, A.I. (1962). *Life, Its Nature, Origin and Development*. Academic Press, New York.

[12] Calvin M. (1969) *Chemical Evolution: Molecular Evolution Towards the Origin of Living Systems on the Earth and Elsewhere*. Oxford University Press, Oxford.

[13] Kamshylov, M.M. (1974) *Evolution of the Biosphere*. (In Russian). Science, Leningrad. Камшилов, М.М. (1974) *Эволюция биосферы*. Наука, Ленинград.

[14] Jukes, T.H. (1966) *Molecules and Evolution*. Columbia University, New York.

[15] Moody, P.A. (1962) *Introduction to Evolution*. Harper, New York.

[16] Hadorn, E., Werner, R. (1989) *General Zoology*. (In Russian). Mir, Moscow. Хадорн, Э., Вернер, Р. (1989) *Общая зоология*. Мир, Москва.

[17] Alberts, B., Bray, D., Lewis, J. et al. (1994) *Molecular Biology of the Cell*, 3rd edn, Vol. 2, Garland Science, New York, p. 461

[18] Fox, S.W., Dose, K. (1977). *Molecular Evolution and the Origin of Life*. M. Dekker, New York.

[19] Nord, F.F., Shubert, W.J. (1962). The biochemistry of lignin formation, in *Comparative Biochemistry: A Comprehensive Treatise,* Vol. IV (eds M. Florkin, and H.S. Mason), Academic Press, New York, p. 62.

[20] Steinman G. *Photobiochemistry*. Ph.D. dissertation, University of California at Berkeley. 1965, p. 146.

[21] Mora, P.T. (1965) The folly of probability. in *The Origins of Prebiological Systems and of Their Molecular Matrices* (ed. S.W. Fox) Academic Press, New York, pp. 310–315.

[22] Alberts, B., Bray, D., Lewis, J. et al. (1994) *Molecular Biology of the Cell*, 3rd edn, Vol. 1, Garland Science, New York, p. 369.

[23] Rutten, M. (1973) *The Origin of Life.* (In Russian), Mir, Moscow. Руттен М. (1973) *Происхождение жизни.* Мир, Москва.

[24] Schopf, J.W., Zeller, O.D. (1976) How old are the eukaryotes? *Science,* **193** (4247), 47–49.

[25] Prosser, C.L. (1970) *Ideas in Evolution and Behavior.* The Natural History Press. Garden City. New York.

[26] Swain, F.M. (1969) Fossile carbohydrates, in *Organic Geochemistry* (eds G. Eglinton, M.T.J. Murphy). Springer–Verlag. Berlin and New York, pp. 374–400.

[27] Schramm, G. (1965) Synthesis of nucleosides and polynucleotides with metaphoric esters, in *The Origins of Prebiological Systems and of Their Molecular Matrices* (ed. S.W. Fox) Academic Press: New York, pp. 309–310.

[28] Hamilton, T.N. (1967) *Process and Pattern in Evolution.* MacMillan, New York.

[29] Browder, L. (1980) *Developmental Biology.* Philadelphia, Saunders.

[30] Needham, A.E. (1965) *The Uniqueness of Biological Materials.* Pergamon. Elmsford, NY.

[31] Sutton-McDowall, M.L., Gilchrist, R.B., Thompson, J.G. (2010). The pivotal role of glucose metabolism in determining oocyte developmental competence. *Reproduction,* **139**, 685–695.

[32] Spicer, A.P., McDonald, J.A. (1988) Characterization and molecular evolution of a vertebrate hyaluronan synthase gene family, *Journal of Biological Chemistry,* **273**, 1923–1932.

[33] Bostock, C.J., Sumner, A.T. (1978) *The Eukaryotic Chromosome,* North-Holland Pub. Co, Elsevier, North Holland.

[34] Alberts, B., Bray, D., Lewis, J. et al. (1994) *Molecular Biology of the Cell*, 3rd edn, Vol. 3, Garland Science, New York, p. 357

[35] Maccari, F., Tripodi, F., Volpi, N. (2004) High-performance capillary electrophoresis of hyaluronan oligosaccharides produced hyaluronate lyase, *Carbohydrate Polymers,* **56**, 55–63.

[36] DeAngelis, P.L., Jing, W., Graves, M.V., Burbanc, D.E. van Etten, J.L. (1997) Hyaluronan synthase of chlorella virus PBCV-1, *Science,* **278**, 1800–1803.

[37] Graves, M.V., Burbanc, D.E., Roth, R., Heuser, J., DeAngelis, P.L., van Etten, J.L. (1999) Hyaluronan synthesis in virus PBCV-1-infected chlorella- like green algae. *Virology,* **257**, 15–23.

[38] Kendall, F.E., Heidelberger, M., Dawson, M.H. (1937) A serologically inactive polysaccharide elaborated by mucoid strains of group A hemolytic *Streptococcus, Journal of Biological Chemistry,* **118**, 61–69.

[39] Carter, G.R., Annau, E. (1953) Isolation of capsular polysaccharides from colonial variants of *Pasteurella multicida. American Journal of Veterinary Research,* **14**, 475–478.

[40] Stepanenko, B.N. (1977) *Chemistry and Biochemistry of Carbohydrates (Polysaccharides).* (In Russian). Vysshaya Shkola, Moscow. Степаненко, Б.Н. (1977) *Химия и биохимия углеводов (полисахариды).* Высшая школа, Москва.

[41] Roberts, S., Urban, J.P.G. (1998) Intervertebral Disks, Chapter 6, Musculoskeletal system, in *Encyclopaedia of Occupational Health and Safety*, (ed. J.M. Stellman), International Labor Office, Geneva.

[42] Polevoy, V.V. (1989) *Plant Physiology.* (In Russian) Vysshaya Shkola, Moscow. Полевой, В.В. (1989) *Физиология растений.* Высшая школа, Москва.

[43] Markovitz, A., Cifonelli, J.A. Dorfman, A. (1959) The biosynthesis of HA by group A *Streptococcus.* 6. Biosynthesis from uridine nucleotides in cell-free extracts. *Journal of Biological Chemistry,* **234** (9), 2343–2350.

[44] DeAngelis, P.L., Weigel, P.H. (1994) Immunochemical confirmation of the primary structure of streptococcal hyaluronan synthase and synthesis of high molecular weight by the recombinant ensime. *Biochemistry,* **33**, 9033–9039.

[45] Alberts, B., Dennis Bray, D., Julian Lewis, J. et al. (1994) *Molecular Biology of the Cell*, 3rd edn, Vol. 4, Garland Science, New York, p. 365.

[46] Shimada, M., Uchida, T. (1998) Immobilisation of hyaluronic acid on egg shell membrane. *Journal of the Society of Fiber Science and Technology,* **44** (2) 108–109.

[47] Stern, R., Asari, A.A., Sugahara, K. (2006) Hyaluronan fragments: an information rich system. *European Journal of Cell Biology*, **85**, 699–715.

[48] Stroitelev, V., Fedorischev, I. (2000) Hyaluronica acid in medicinal and cosmetic preparations (in Russian). *Kosmetika i medicina*, **3**, 21–34. Строителев В., Федорищев И. (2000) Гиалуроновая кислота в медицинских и косметических препаратах. *Косметика и медицина*, **3**, 21–34.

[49] Kielty, C.M., Whittaker, S.P., Grant, M.E., Shuttleworth, C.A. (1992) Type IV collagen microfibrils: evidence for structural association with hyaluronan. *Journal of Cell Biology*, **118** (4), 979–990.

[50] Pianigiani, E., Andressi, A., Taddenci, P. (1999) A new model for studying differentiation and growth of epidermal cultures of hyaluronan-based carrier. *Biomaterials*, **20** (18), 1689–1694.

[51] Culty, M., Nguyen, H.A., Underhill, C.B. (1992) The hyaluronan receptor CD44 participates in uptake and degradation of hyaluronan. *Journal of Cell Biology*, **116**, 1055–1062.

[52] Gurdon, J. (1974) *The Control of Gene Expression in Animal Development.* Oxford University Press, Oxford.

[53] Ham, R.G, Veomett, M.I. (1979) *Mechanisms of Development.* Mosby, St. Louis, MI.

[54] Wessels, N.K. (1977) *Tissue Interactions and Development.* Benjamin Cummings, Menlo Park CA.

[55] Trinkaus, J.P. (1969) *Cells into Organs: The Forces That Shape the Embryo.* Prentice Hall, Upper Saddle River, NJ.

[56] Yung, S., Thomas, G.J., Davies M. (2000). Induction of hyaluronan metabolism after mechanical injury of human peritoneal mesothelial cells in vitro. *Kidney International*, **58,** 1953–1962.

[57] Slutsky, L.I. (1969) *Biochemistry of normal and pathological connective tissue.* (In Russian) Medicina, Leningrad. Слуцкий, Л.И. (1969) *Биохимия нормальной и патологически измененной соединительной ткани.* Медицина, Ленинград.

[58] Campo, R.D., Tourtellotte, C.D. (1967) The composition of bovine cartilage and bone. *Biochimica et Biophysica Acta*, **141** (3), 614–624.

[59] Chvapil, M. (1968) *Physiology of Connective Tissue.* Butterworth, London and Czechoslovac Medical Press, Prague.

[60] Lash, J.W., Saxen, L., Kosher, R.A. (1974) Human chondrogenesis: glycosaminoglycan content of embryonic human cartilage. *Journal of Experimental Zoology*, **189** (1), 127–131.

[61] Kondo, K., Seno, N., Anno, K. (1971) Mucopolysaccharides from chick of three age groups. *Biochimica et Biophysica Acta*, **244** (3), 513–522.

[62] Lipson, M.J. Cerskus, R.A., Silbert, J.E. (1971) Glycosaminoglycans and glycosaminoglycan-degrading ensime of Rana cayesbeiana back skin during late stages of metamorphosis. *Developmental Biology*, **25** (2), 198–208.

[63] Meaney, M.F. (1974) Increased accuracy and precision in screening for urinary mucopolisaccharides. *Journal of Medical Laboratory Technology*, **28** (2), 29–34.

[64] Nikitin, V.N., Persky, E.E., Utevskaya, L.A. (1977) *Age and evolutionary biochemistry of collagen structures.* (In Russian) Naukova Dumka, Kiev. Никитин, В.Н., Перский, Е.Э., Утевская, Л.А. (1977) *Возрастная и эволюционная биохимия коллагеновых структур.* Наукова Думка, Киев.

[65] Adhirajan, N., Shanmugasundaram, N., Shanmuganathan, S., Babu, M. (2009) Collagen-based wound dressing for doxycycline delivery: *in-vivo* evaluation in an infected excisional wound model in rats. *Journal of Pharmacy and Pharmacology*, **61**, 1617–1623.

[66] Kelly, G.S. (1998) The role glucosamine sulfate and chondroitin sulfate in the treatment of degenerative joint disease. *Alternative Medicine Review*, **3** (1), 27–39.

[67] Setnikar, I., Pacini, M.A., Revel, L. (1991) Antiarthritic effect of glucosamine sulfate studied in animal models. *Arzneimittel Forschung*, **41** 542–545.

[68] Setnikar, I. (1992) Antireactive properties of chondroprotective drugs. *International Journal of Tissue Reactions*, **41**, 253–261.

[69] Andrews, J.L., Ghosh, P., Lentini, A., Ternal, B. (1983) The interaction of pentosan polysulphate with human neutrophyl elastase and connective tissue matrix components. *Chemico-Biological Interactions*, **47**, 157–173.

[70] Baici, A., Salgam, P., Fehr, K., Boni A. (1980) Inhibition of human elastase from PMN leucocytes by glycosaminoglycan polysulphate (arteparon). *Biochemical Pharmacology*, **29**, 1723–1727.

[71] Halverson, P.B., Cheung, H.S., Struve, J., McCarty, D.J. (1987) Suppression of active collagenase from calcifid lapine synovium by arteparon. *Journal of Rheumatology,* **14**, 1013–1017.

[72] Carreno, M.R., Muniz, O.E., Howell, D.S. (1986) The effect of glycosaminoglycan polysulfuric acid ester on articular cartilage in experimental osteoarthritis: effect on morphological variables of disease severity. *Journal of Rheumatology,* **13**, 490–497.

[73] Kontrawelert, P., Francis, D.L., Brooks, P.M., Ghosh, P. (1989) Application of an enzyme-linked immunosorbent-inhibition assay to quantitate the release of KS peptides into fluids of the rat subcutaneous air pouch model and the effects of chondroprotective drugs on the release process. *Rheumatology International,* **9** (2), 77–83.

[74] Hannan, H., Ghosh, P., Bellenger, C., Taylor, T. (1987) Systemic administration of glycosaminoglycan polysulphate (arteparon) provides partial protection of articular cartilage from damage produced by meniscectomy in the canine. *Journal of Orthopaedic Research,* **5** (1), 47–59.

[75] Postmen, O., Forsskahl, B., Markhind, M. (1988) Local glycosaminoglycan polysulphate injection therapy in osteoarthritis of the hand. A placebo controlled double-blind clinical study. *Scandinavian Journal of Rheumatology,* **17**, 197–202.

[76] Setnikar, I., Pacini, M.A., Revel, L. (1991) Antiarthritic effect of glucosamine sulfate studied in animal models. *Arzneimittel Forschung,* **41**, 542–545.

[77] Conrozier, T. (1998) Death of articular chondrocytes. Mechanisms and protection. *La Presse Médicale,* **27** (36), 1859–1861.

[78] Deal, C.L., Moskowitz, R.W. (1999) Nutraceuticals as therapeutic agents in osteoarthritis. The role of glucosamine, chondroitin sulfate and hydrolisate. *Rheumatic Disease Clinics of North America,* **25** (2), 379–395.

[79] Van der Kraan, P.M., de Vries, B.J., Vitters, E.L. (1988) Inhibition of glycosaminoglycan synthesis in anatomically intact rat patellar by paracetamol-induced serum sulfate depletion. *Biochemical Pharmacology,* **37** (6), 3683–3690.

[80] Van der Kraan, P.M., Vitters, E.L., de Vries, B.J., van den Berg, W.B. (1990) High susceptibility of human articular cartilage glycosaminoglycan synthesis to changes in inorganic sulfate availability. *Journal of Orthopaedic Research,* **8** (4), 565–571.

[81] Noden, D.M. (1978) Interaction directing migration and cytodifferentiation of avian neural crest cells, in *Specificity of Embryological Interactions (Receptors and Recognition, Series B, Vol. 4)* (ed. D.R. Garrod). Chapman and Hall, London, p. 217.

[82] Tosney, K.W. (1982) The segregation and early migration of cranial neural crest cells in the avian embryo. *Developmental Biology,* **89**, 13–24.

[83] Crienberg, J.H., Seppa, S., Seppa, H., Hewitt, A.T. (1981) Role of collagen and fibronectin in neural crest adhesion and migration. *Developmental Biology,* **87**, 259–266.

[84] Le Douarin, N.M. (1980) The ontogeny of neural crest in avian embryo chimaeras. *Nature,* **286**, 663–669.

[85] Patterson, P.H. (1978) Environmental determination of autonomic neurotransmitter function. *Annual Review of Neuroscience,* **1**, 1–17.

[86] Coon, H.G. (1966) Clonal stability and phenotypic expression of chick cartilage cells in vitro. *Proceedings of the National Academy of Sciences of the USA,* **55**, 66–73.

[87] Von der Mark, K., Gauss, V., Von der Mark, H., Muller, P. (1977) Relationship between cell shape and type of collagen synthesized as chondrocytes lose their cartilage phenotype in culture. *Nature,* **267**, 531–532.

[88] Archer C.W., Rooney P., Wolpert L. (1982) Cell shape and cartilage differentiation of early chick limb bud cells in culture. *Cell Differentiation,* **11** (4), 245–251.

[89] Bullingham, R.E., Silvers, W.K. (1967) Studies on the conservation of epidermal specificities of skin and certain mucosas in adult mammals. *Journal of Experimental Medicine,* **125**, 429–446.

[90] Eguchi, G. (2008) 'Transdifferentiation' of vertebrate cells in cell culture, in *Ciba Foundation Symposium 40 – Embryogenesis in Mammals* (eds K. Elliott, M. O'Connor), John Wiley & Sons, Ltd, Chichester, pp. 241–257.

[91] Razin, A., Riggs, A.D. (1980) DNA methylation and gene function. *Science,* **210**, 604–610.

[92] Jones, P.A., Taylor, S.M. (1980) Cellular differentiation, cytidine analogs and DNA methylation, *Cell,* **20**, 85–93.

[93] Felsenfeld, G., McGhee, J. (1982) Methylation and gene control. *Nature,* **296**, 602–603.

[94] Alexandrov, V.Y. (1981) *Reactivity of Cells and Proteins.* (In Russian), Nauka, Leningrad. Александров В.Я. (1981) *Реактивность клеток и белки.* Наука, Ленинград.

[95] Brodsky, V.Y., Uryvaeva I.V. (1981) *Cell Polyploidy. Proliferation and Differentiation.* (In Russian), Nauka, Moscow. Бродский, В.Я., Урываева, И.В. (1981) *Клеточная полиплоидия. Пролиферация и дифференцировка.* Наука, Москва.

[96] Brodsky, V.Y., Nechayev, N.V. (1988) *Protein Synthesis Rhythm,* (In Russian), Nauka, Moscow. Бродский В.Я., Нечаева Н.В. (1988) *Ритм синтеза белка.* Наука, Москва.

[97] Marzullo, G., Lash, J.W. (1970) Control of phenotypic expression in cultured chondrocytes: investigation on the mechanism, *Developmental Biology,* **22**, 638–654.

[98] Goss, R.J. (1978) *The Physiology of Growth,* Academic Press, New York.

[99] Holder, N. (1981) Regeneration and compensatory growth. *British Medical Bulletin,* **37**, 227–232.

[100] Grenader, A.K. (1984) A possible mechanism of stimulation of cell proliferation by various damaging effects (in Russian). *Biofizika,* **29** (5), 840–841. Гренадер, А.К. (1984) О возможном механизме стимулирования пролиферации клеток различными повреждающими воздействиями. *Биофизика,* **29** (5), 840–841.

[101] Vasiliev, J.M., Gelfand, I.M., Gelstein, V.I. (1971) Initiation of DNA synthesis in cultures of mouse fibroblast-like cells by the action of the substances that disrupt the formation of microtubules (in Russian). *Doklady Akademii Nauk SSSR,* **197**, 1425–1428. Васильев, Ю.М., Гельфанд, И.М., Гельштейн, В.И. Инициация синтеза ДНК в культурах мышиных фибробластоподобных клетках при действии веществ, нарушающих формирование микротрубочек. *Доклады Академии Наук СССР,* **197**, 1425–1428.

[102] Cohn, S.M., Krawics, B.R., Drebler, S.L., Lieberman, M.W. (1984) Induction of replicative DNA synthesis in quiescent human fibroblast by DNA damaging agents. *Proceedings of the National Academy of Sciences of the USA,* **81** (15), 4828–4832.

[103] Kaz'min, S.D. (1984) *Biochemistry of the mitotic cycle of tumor cells* (In Russian). Naukova Dumka, Kiev. Казьмин С.Д. *Биохимия митотического цикла опухолевых клеток.* Наукова Думка, Киев.

[104] Todorov, I. Boĭkov, P.Y., Sidorenko, L. et al. (1978) Stimulation of DNA replication in rat liver cells as a result of inhibition of protein synthesis (in Russian). *Doklady Akademii Nauk SSSR,* **239** (5), 1255–1258. Тодоров, И.И., Бойков, П.Я., Сидоренко, Л. И. и др. (1978) Стимуляция репликации ДНК в клетках печени крыс как результат ингибирования синтеза белков. *Доклады Академии Наук СССР,* **239** (5), 1255–1258.

[105] Ponta, H., Sherman, L., Herrlich, P.A. (2003) CD44: From adhesion molecules to signaling regulators. *Nature Reviews Molecular Cell Biology,* **4** (1), 33–45.

[106] Xu, H., Ito, T., Tawada, A. et al. (2002) Effect of hyaluronan oligosaccharides on expression of heat shock protein 72. *Journal of Biological Chemistry,* **277**, 17308–17314.

[107] West, D.C. (1985) Angiogenesis induced by degradation products of hyaluronic acid, *Science,* **228**, 1324–1326.

[108] Boĭkov, P.Y., Sidorenko, L.I, Shevchenko, N.A., Todorov I.N. (1981) Biogenesis of chromatin in the cells of higher animals. Acceleration transport non-histone proteins in the nucleus and their catabolism in conditions inhibit protein synthesis (in Russian). *Biokhimiia,* **46** (8), 1396–1410. Бойков П.Я., Сидоренко Л.И., Шевченко Н.А., Тодоров И.Н. (1981) Биогенез хроматина в клетках высших животных. Ускорение транспорта негистоновых белков в ядро и их катаболизма в условиях ингибироваия синтеза белков. *Биохимия.,* **46** (8), 1396–1410.

[109] Gutnikova, M.N., Shevchenko, N.A., Boĭkov, P.Y. et al. (1985) Activation of nuclear polypeptide synthesis in the liver cells of rats during suppression of matrix protein synthesis. The contribution of nuclear polypeptide synthesis into changes in chromatin structures (in Russian). *Biokhimiia,* **50** (10), 1990– 1995. Гутникова, М.Н., Шевченко, Н.А., Бойков, П.Я.

и др. (1985) Активация ядерного полипептидного синтеза в клетках печени крыс при торможении матричного синтеза белков. Вклад ядерного полипептидного синтеза в изменения структур хроматина. *Биохимия*, **50** (10), 1990–1995.

[110] Shevchenko, N.A., Boĭkov, P.Y., Todorov, I.N. (1985) Sequential changes in macrostructure state of chromatin and RNA-synthesizing activity of nucleolus and extra nucleolus chromatin of rat liver cells in the induction of DNA synthesis (In Russian). *Biokhimiia*, **50** (10), 1591–1598. Шевченко, Н.А., Бойков, П.Я., Тодоров И.Н. (1985) Последовательные изменения макроструктурого состояния хроматина и РНК-синтезируюей активности ядрышкового и экстраядрышкового хроматина клеток печени крыс в процессе индукции синтеза ДНК. *Биохимия*, **50** (10), 1591–1598.

[111] Boĭkov, P.Y., Sidorenko, L.I., Todorov, I.N. (1979) Biogenesis of chromatin in the cells of the highest animals. Activation of synthesis of nuclear proteins and hepatocyte DNA after impulse inhibition of translation by cycloheximide (in Russian). *Biokhimiia*, **44** (6), 963–974. Бойков, П.Я., Сидоренко, Л.И., Тодоров, И.Н. (1979) Биогенез хроматина в клетках высших животных. Активация синтеза ядерных белков и ДНК гепатоцитов после импульсного торможения трансляции циклогексимидом. *Биохимия*, **44** (6), 963–974.

[112] Borisova, N.P., Kostyuk, G.V., Shevchenko, N.A., et al. (2003) Endogenous DNases as a tool for isolation of nuclear matrix: critical parameters of nucleolysis. *Biology Bulletin of the Russian Academy of Sciences,* **30** (5), 442–448.

[113] Boĭkov, P.Y. (1987) Mechanisms of induction of differentiated cells proliferation (in Russian). PhD Thesis, Moscow State University, Moscow, p. 587. Бойков П.Я. Механизмы инициирования пролиферации дифференцированных клеток. Дисс. На соиск. д.б.н. М.- 1987.-С. 587.

[114] Borisova, N.P., Kostyuk, G.V., Shevchenko, N.A. et al. (2003) Functional association of immediate early gene c-fos with nuclear matrix. *Bulletin of Experimental Biology and Medicine,* **135** (2), 167–170.

[115] Boĭkov, P.Y., Shevchenko, N.A., Sidorenko, L.I. (1998) Dynamics of association-dissociation of proto-oncogene c-myc with the nuclear matrix during the activation-inactivation of gene (In Russian). *Biokhimiia,* **63** (5), 631–638. Бойков, П.Я., Шевченко, Н.А. Сидоренко, Л.И. (1998) Динамика ассоциации-диссоциации протоонкогена c-myc с ядерным матриксом в процессе активации-инактивации гена. *Биохимия,* **63** (5), 631–638.

[116] Boĭkov, P.Y., Shevchenko N.A., Sidorenko L.I., Todorov, I.N. (1984) Ratio of biosynthesis of intracellular and export proteins in rat liver cells during the induction of proliferation by cyclohexamide (In Russian). *Biokhimiia.* **49** (9), 1470–1477. Бойков П.Я., Шевченко Н.А., Сидоренко Л.И. Тодоров И.Н. (1984) Соотношение биосинтеза внутриклеточных и экспортных белков в клетках печени крыс в процессе индукции пролиферации циклогексимидом. *Биохимия,* **49** (9), 1470–1477.

[117] Green, H. (1977) Terminal differentiation of cultured human epidermal cells. *Cell,* **11**, 405–415.

[118] Alessenko, A.V., Boĭkov, P.Y., Filippova, G.Y., et al. (1997) Mechanisms of cyclohexamide-induced apoptosis in liver cells. *FEBS Letters,* **416**, 113–116.

[119] Sengel, P. (1976) *Morphogenesis of Skin,* Cambridge University Press, Cambridge.

[120] Fuchs, E., Green, H. (1980) Changes in keratin gene expression during terminal differentiation of the keratinocyte. *Cell,* **19**, 1033–1042.

[121] Dasgeb, B., Phillips, T. (200) What are scars?, in *Scar Revision,* 1st edn, (ed. K.A. Arndt), Elsevier/Saunders, Philadelphia, PA.

[122] Astériou, T., Deschrevel, B., Gouley F., Vincent, J.C. (2002) Influence of substrate and enzyme concentration on hyaluronan hydrolysis kinetics catalyzed by hyaluronidase, in *Hyaluronan: Proceedings of an International Meeting, September 2000, North East Wales Institute, UK* (eds J.F. Kennedy, G.O. Phillips, P.A. Williams, V.C. Hascall), Woodhead Publishing Ltd, Cambridge, pp. 249–252.

[123] Underhill C.B. (1989) The interaction of hyaluronate with the cell surface: the hyaluronate receptor and the core protein, in *Ciba Foundation Symposium 143 – The Biology of Hyaluronan* (eds D. Evered, J. Whelan), John Wiley & Sons, Ltd, Chichester, pp. 60–86.

[124] Comper, W.D., Laurent, T.G. (1978) Physiological function of connective tissue polysaccharides. *Physiological Reviews,* **58**, 255–315.

[125] Weigel, P.H., Frost, S.J., Leboeuf, R.D., McGary, C.T. (2007) The specific interaction between fibrinogen and hyaluronan: possible consequences in haemostasis, inflammation and wound healing in *Ciba Foundation Symposium 143 – The Biology of Hyaluronan* (eds D. Evered, J. Whelan), John Wiley & Sons, Ltd, Chichester, pp 248–264.

[126] Tammi, M.I., Day, A.J., Turley, E.A. (2002) Hyaluronan and homeostasis: a balancing act. *Journal of Biological Chemistry,* **277**, 4581–4784.

[127] Goodwin, B.C. (1963) *Temporal Organization in Cells. A Dynamic Theory of Cellular Control Processes.* Academic Press, London, New York.

[128] West, D.C., Hampson, I.N., Arnold, F., Kumar S. (1985) Angiogenesis induced by degradation products of hyaluronic acid. *Science,* **228**, 1324–1326.

[129] Delpech, B., Girard, N., Bertrand, P., Courel, M.N. (1997) Hyaluronan: fundamental principles and applications in cancer. *Journal of Internal Medicine,* **7**, 41–48.

[130] Eliseyev, V.G. (1961) *Connective tissue.* (In Russian), Medgiz, Moscow.Елисеев, В.Г. (1961) *Соединительная ткань.* Медгиз, Москва.

[131] Konyshev, V.A. (1974) *Stimulators and Inhibitors of the Animals Organs and Tissues.* (In Russian), Medicine, Moscow. Конышев, В.А. (1974) *Стимуляторы и ингибиторы органов и таней животных.* Медицина, Москва.

[132] Kruglikov, G.G., Arutyunov, V.D., Batsura, Y.D., Shimkevich, L.L. (1977) The differentiation of fibroblasts in the collagen formation (in Russian). *Ontogenesis,* **2**, 186–190. Кругликов, Г.Г., Арутюнов, В.Д., Бацура, Ю.Д., Шимкевич, Л.Л., (1977) Дифференцировка фибробластов в процессе коллагенообразования. *Онтогенез,* **2**, 186–190.

[133] Serov, V.V., Schechter, A.B. (1981) *Connective Tissue.* (In Russian), Medicina, Moscow. Серов, В.В., Шехтер, А.Б. (1981) *Соединительная Ткань.* Медицина, Москва

[134] Yakovleva, A.A. (1992) Excretion of hyaluronic acid and chondroitin rheumatoid arthritis (In Russian). *Reumatologiia* **1**, 5–8. Яковлева, А.А. (1992) Экскреция гиалуроновой кислоты и хондроитинсульфатов при ревматоидном артрите. *Ревматология,* **1**, 5–8.

[135] Costa, E., Trabuchi, M. (eds) (1978) *The Endorphins.* Raven Press, New York.

[136] Atkins, E.D., Shechan, J.K. (1973) Hyaluronans: Relation between molecular conformations. *Science,* **179** (4073), 562–564.

[137] Scott, J. (2007) Secondary structures in hyaluronan solutions: chemical and biological implications, in *Ciba Foundation Symposium 143 – The Biology of Hyaluronan* (eds D. Evered and J. Whelan), John Wiley & Sons, Ltd, Chichester, pp. 6–20.

[138] Onishchenko, K. (2008) The extracellular matrix (in Russian). *Esteticheskaiia Medicina,* **7** (4), 449–456. Онищенко, К. (2008) Внеклеточный матрикс. *Эстетическая медицина,* **7** (4), 449–456.

[139] Kogan, G., Soltes, L., Stern, R., et al. (2008) Hyaluronic acid: its function and degradation in *in vivo* systems, in *Studies in Natural Products Chemistry,* vol. 34 (ed. Atta-ur-Rahman), Elsevier, Amsterdam, pp. 131–143.

[140] Brown, G.H., Wolken, J.J. (1979) *Liquid Crystals and Biological Structures.* Academic Press, New York, San Francisco, London.

[141] Tal'roze, R.V, Shatalova, A.M., Shandryuk, G.A. (2009) Development and stabilization of liquid crystalline phases in hydrogen-bonded systems. *Polymer Sciences. Series B,* **51** (3), 489–516.

[142] Evdokimov, Y.M. (2008) *Liquid Crystal Dispersions and DNA Nanoconstruction.* (In Russian). Radiotechnika, Moscow. Евдокимов, Ю.М. *Жидкокристаллические дисперсии и наноконструкции ДНК.* Радиотехника, Москва.

[143] Shoseyov, O., Levy, I. (eds) (2008) *Nanobiotechnology. Bioinspired Devices and Material of the Future,* Humana Press, Totowa, NJ.

[144] Budihina, A.S, Pinegin, B.V. (2008) Alpha defensins — antimicrobial peptides of neutrophil: properties and function (in Russian). *Russian Journal of Immunology,* **29** (5), 317–320.

[145] Gamazkov, O.A. (1995) *Physiologically Active Peptides: Reference Manager* (in Russian). IPGM, Moscow. Гамазков О.А. *Физиологически активные пептиды: справочное руководство.* М.: ИПГМ.- 1995.-С. 296.

[146] Lysenko, A.V., Harutyunyan, A.V., Kozin, L.S. (2005) *Peptide Regulation of Body Adaptation to Stress Actions.* (In Russian). Voenno-Medicinskaia Akademiia, St. Petersburg. Лысенко, А.В., Арутюнян, А.В., Козина, Л.С. (2005) *Пептидная регуляция адаптации организма к стрессорным воздействиям.* Военно-медицинская академия, Санкт-Петербург.

[147] Kozina, L.S., Stvolinsky, S.A., Fedorov, T.N. et al. (2008) Study of antioxidant properties of short peptides in model experiments (In Russian). *Problems of Biological, Medical and Pharmaceutical Chemistry.* **6**, 31–36. Козина, Л.С., Стволинский, С.А., Федорова, Т.Н. и др. (2008) Изучение антиоксидантных свойств коротких пептидов в модельных экспериментах. *Вопросы биологической, медицинской и фармацевтической химии,* **6**, 31–36.

[148] Bradshaw, R.A., Dennis E.A. (eds) (2009) *Handbook of Cell Signaling,* Elsevier, Amsterdam.

[149] Ashmarin, I.P. (2005) Hormones and regulatory peptides: the differences and similarities of concepts and functions. The place of hormones among other intercellular signaling. *Russian Chemical Journal,* **1**, 4–7 (In Russian). Ашмарин, И.П. (2005) Гормоны и регуляторные пептиды: различия и сходства понятий и функций. Место гормонов среди других межклеточных сигнализаторов. *Российский химический журнал,* **1**, 4–7.

[150] Austel, C.C., Nure, E.A., Vandelig, K. (1991) Hyaluronan and cell-associated hyaluronan binding protein regulate the locomotion of ras-transformid cells. *Journal of Cell Biology,* **112** (5), 1041–1047.

[151] Ocamoto, I. (2001) Proteolitic release of CD44 tetracellular domain and its role in the CD44 signaling pathway. *Journal of Cell Biology,* **155**, 755–762.

[152] Pure, E., Cuff, C.A. (2001) A crucial role for CD44 in inflammation. *Trends in Molecular Medicine,* **7,** 313–221.

[153] Paston, J.H., Willingham, M.C. (1981) Receptor-mediated endjcytosis of hormonesin cultured cells. *Annual Review of Physiology,* **43**, 239–250.

[154] Kahn, C.R. (1976) Membrane receptors for hormones and neurotransmitters. *Journal of Cell Biology,* **70**, 260–286.

[155] Sutherland E.W. (1972) Studies on the mechanism of hormone action. *Science,* **177**, 401–408.

[156] Hoppe, J., Wagner, K.G. (1979) Cycle AMP-dependent protein kinase I, a unique allosteric enzyme. *Trends in Biochemical Sciences,* **4**, 282–285.

[157] Smith, S.B., White, H.D., Siegel, J.B., Krebs, E.G. (1981) Cyclic AMP-dependent proteinkinase I: cyclic nucleotide binding, structural changes, and release of the catalytic subunits. *Proceedings of the National Academy of Sciences of the USA,* **78**, 1591–1595.

[158] Krebs, E.G., Beavo, J.A. (1979) Phosphorylation-dephosphorylation of enzymes. *Annual Review of Biochemistry,* **48**, 923–959.

[159] Means, A.R., Dedman, J.R. (1980) Calmodulin – an intracellular calcium receptor. *Nature,* **285**, 73–78.

[160] Kretsinger, R.H. (1981) Mechanisms of selective signaling by calcium. *Neuroscience Research Program Bulletin,* **19**, 213–328.

[161] Stern, R., Asari, A.A., Sugahara, K.H. (2006) Hyaluronan fragments: an information-rich system. *European Journal of Cell Biology,* **85**, 699–715.

[162] Hay, E.D. (1981) Extracellular matrix. *Journal of Cell Biology,* **91**, 205–223.

[163] Hay, E.D. (1982) *Cell Biology of Extracellular Matrix.* Plenum Press, New York.

[164] Laurent, T.C. (1970) Structure of hyaluronic acid, in *Chemistry and the Molecular Biology of the Intracellular Matrix.* (ed. E.A. Balazs). Academic Press, London, pp. 703–732.

[165] Lindhall, U., Hook. M., (1978) Glycosaminoglycans and their binding to biological macro-molecules. *Annual Review of Biochemistry,* **47**, 385–417.

[166] Lepperdinger, G., Fehrer, C., Reitinger, S. (2004) Biodegradation of Hyaluronan, in *Chemistry and Biology of Hyaluronan* (eds H.G. Garg and C.A. Hales), Elsevier, Amsterdam, pp. 71–82.

[167] Heinegard, D., Bjornsson, S., Morgelin, M., Sommarin, Y. (1998) Hyaluronan-binding matrix proteins, in, *The Chemistry, Biology and Medical Applications of Hyaluronan and Its Derivatives* (ed. T.C. Laurent), Portland Press, London.

[168] Schmidt, R.F., Thews, G. (eds), (1989) *Human Physiology.* Verlag, Berlin-Heidelberg/New York.

[169] Musil, J., Novakova O., Kunz K. (1980) *Biochemistry in Schematic Perspective.* Czechoslovac Medical Press, Prague.

[170] Vance D.E., Vance J.E. (2008) *Biochemistry of Lipids, Lipoproteins, and Membranes.* Elsevier, New York, p. 361.

[171] Bezuglov, V.V., Konovalov, S.S. (2009) *Lipids and Cancer,* Prime Evroznak. Sankt Petersburg (in Russian). Безуглов, В.В., Коновалов, С.С. (2009) *Липиды и рак,* Прайм-Еврознак. Санкт-Петербург.

[172] Paston, J.H., Willingham, M.C. (1981) Receptor-mediated endocytosis of hormones in cultured cells. *Annual Review of Physiology,* **43**, 239–250.

[173] Kahn, C.R. (1976) Membrane receptors for hormones and neurotransmitters. *Journal of Cell Biology,* **70**, 261–286.

[174] Smith, S.B., White, H.D., Siegel, J.B., Krebs E.G. (1981) Cyclic AMP-dependent proteinkinase 1: cyclic nucleotide binding, structural changes, and release of the catalytic subunits. *Proceedings of the National Academy of Sciences of the USA,* **78**, 1591–1595.

[175] Maytin, E.W., Chung, H.H., Seetharaman, V.M. (2004) Hyaluronan participates in the epidermal response to disruption of the permeability barrier *in vivo. American Journal of Pathology,* **165** (4), 1331–1341.

[176] Zubarov, D.M., Zubarova L.D. (2009) *Microvesicles in Blood.* GEOTAR-MEDIA, Moscow. Зубаров, Д.М., Зубарова, Л.Д. (2009) *Микровезикулы в крови.* ГЭОТАР-МЕДИА, Москва.

[177] Stern, R., Frost, G.J., Shuster, S. et al. (1998) Hyaluronic acid and skin. *Cosmetics & Toiletries,* **113**, 43–48.

[178] Manuskatti, W., Maibach, H.J. (1996) Hyaluronic acid and skin: wound heating and aging. *International Journal of Dermatology,* **35** (8), 539–544.

[179] Karvinen, S., Pasonen-Seppanen, S., Hyltinen, J. (2003) Keratinocyte growth factor stimulates migration and hyaluronan synthesis in the epidermis by activation of keratinocyte hyaluronan synthases 2 and 3. *Journal of Biological Chemistry,* **278**, (49), 49495–49504.

[180] Nikolaev, S.S., Bykov, V.A., Yakovleva, L.V. et al. (2002) Biochemical and moisture-exchange characteristics of the surface layer of articular cartilage. (in Russian). *Bulletin of Experimental Biology and Medicine,* **133** (10), 390–397. Николаева, С.С., Быков, В.А., Яковлева, Л.В., и др. (2002) Биохимические и влагообменные характеристики поверхностного слоя суставного хряща. *Бюллетень экспериментальной биологии и медицины,* **133** (10), 390–397.

3

Methods of Hyaluronic Acid Production

Upon first glance, it may appear this chapter is oversaturated with technological details that would be of interest only to technologists working in hyaluronan production. However, the material is presented in this way intentionally in order to properly elucidate the difficulties of hyaluronic acid production and the potential problems caused by the technological impurities in the final product when intended for medical application.

3.1 Hyaluronan Sources and Extraction

Hyaluronan can be extracted from the tissue of vertebrate animals (Table 3.1) or from the bacteria that creates the protecting capsule from the polysaccharides. The best bacterial strain for such a purpose is the haemolytic *Streptococcus* groups A and B.

3.1.1 Hyaluronan Production from Animal Sources: General Methods

Until recently, the most economically viable way of obtaining HA was by its extraction from chicken combs [1,2]. A full procedure includes subsequent stages of tissue homogenization, extraction, purification and preparation of the final commercial HA product, which can be in the form of dry powder, solution, granules or in medicinal substances. The novelty of the various patent-protected methods is that they each involve different extraction technologies and/or purification procedures. Next is the typical procedure for the the extraction and purification of hyaluronan from chicken combs.

Rooster combs contain up to 7.5 mg of the polysaccharide per 1 g of tissue. The polysaccharide is mainly localized in the mucous fibres of the subcutaneous layer. Before extraction from the freshly frozen tissue, the combs must be washed thoroughly with water, acetone

Hyaluronic Acid: Preparation, Properties, Application in Biology and Medicine, First Edition.
Mikhail A. Selyanin, Petr Ya. Boykov and Vladimir N. Khabarov.
© 2015 John Wiley & Sons, Ltd. Published 2015 by John Wiley & Sons, Ltd.

Table 3.1 *Hyaluronic acid content in various tissues*

Tissue (liquid)	Content, mg/g (ml)
Adult roster comb	Up to 7.5
Eye vitreous body	0.1–0.4
Synovial liquid	1.3–4.0
Hyaline cartilage	1.0
Umbilical cord	4.1
Epidermis	0.1
Derma	0.2–0.5
Amniotic fluid	20

(5–10 extractions till transparent solution) [3], 95% ethanol [4] or a mixture of ethanol and chloroform. This procedure is necessary in order to avoid enzymatic and oxidative destruction of HA, which has a free-radical mechanism with the involvement of ions of iron, copper and phosphate [1].

Blood-free tissues could be stored for up to 24 months in 95% ethanol at 4–22°C [3]. Then the tissues should be grounded in a homogenizer, disintegrator or ball mill in order to afford maximum extraction. Different solvents could be used for extraction, including distilled water at different temperatures, salt solutions and aqueous-organic mixtures. Several of the most common methods of the production of hyaluronic acid from the rooster comb are presented in the Table 3.2, although one method describes production from the eggshell membrane.

Up to 93% of a chicken comb's hyaluronic acid can be obtained when water is extracted at a tissue temperature of 80–100ºC. At the same time, an inactivation of hyaluronidases takes place. At high temperatures, however, HA undergoes partial depolymerization regardless of the enzymatic hydrolysis. It is therefore necessary to find the optimal temperature regime in order to ensure inactivation of the decomposition enzymes by the high temperature and avoid (or at least minimize) biopolymer destruction.

A method for hyaluronic acid extraction by aqueous-alcohol mixture, containing 5–25% n-propyl, iso-propyl or *tert*-butyl alcohol, 2 M calcium chloride (of cartilage origin) is described. In order to isolate hyaluronan from its complexes with proteins, a homogenized tissue is treated with proteolytic enzymes (pepsin, trypsin, papain and/or pronase) and aqueous-organic mixtures. But in most methods these procedures are included in the treatment of the primary extracts.

3.1.2 Hyaluronan Purification

After extraction, hyaluronan must undergo purification as the next production operation. The extract from the animal tissue usually contains the following impurities: proteins, peptides, lipids, nucleic acids, mucopolysacchrides and low molecular weight precursors. The first purification stage involves the precipitation of HA from the primary extract using ethanol or acetone or acetic acid, or a double volume of ethanol with sodium acetate at 2–4°C [4]. Sometimes the dissolution-precipitation cycles are repeated several times in order to help remove low-molecular weight compounds and lipids that are soluble in acetone and ethanol. The proteins (which are free and connected with the polysaccharides) are removed

#	Source	Extraction	Purification	Final Product	Reference
1	Rooster combs	Water 100°C, 6 times	Papain; ultrafiltration in 40% water-ethanol mixture.	Lyophilized powder of sodium hyaluronan	[5]
2	Rooster combs	Water	Extract heating at 90–100°C; lipid removal; filtration; treatment with activated carbon.	Lyophilized powder of hyaluronan	[6]
3	Rooster and chicken combs	Water acidified to pH 3–4, 90–100°C, 40–50 min.	Treatment with activated carbon then cellulose; filtration.	Lyophilized powder of hyaluronan, protein content below 0,05%	[7]
4	Rooster and Chicken Combs	Water, 2 extractions	Treatment with chloroform; precipitation with ethanol.	Lyophilized powder of hyaluronan, protein content below 0.5%	[8]
5	Chicken combs	Aqueous solution of n-propyl or tert-butyl alcohol twice (5 – 25%) liquid module 1:(10–15)	Sodium chloride to creation of two-phase system; precipitation with ethanol.	White amorphous powder, protein content below 1%	[9]
6	Rooster combs	Physiological solution, 80–90°C, 2 extractions	Filtration; precipitation with saponified acetic acid with sodium hydroxide to pH 7.0–7.3; heating to 80–90°C; repeatable filtration.	Lyophilized powder free from nucleic acids	[10]
7	Rooster combs	Water extraction	Multiple treatments with a mixture chloroform and sodium chloride 4–5°C for 3–5h; treatment with pronaze; precipitation with ethanol.	'Healon'	[4]
8	Rooster combs	Water, 3 extractions. Tissue – water 1(4–6), 2–4 h	Precipitation with trichloroacetic acid (1 – 2%) from the extract volume at 20–22°C for 1–2 h; lipid and water removal with acetone and ether three times.	Lyophilized powder	[11]
9	Rooster combs	1–15% solution of sodium chloride at 60°C, 18 h. Yield 1.92% from the stating material.	Centrifugation; lyophilization.	Fibre-like white substance; protein content 9–24%; optical density of 0.1% aqueous solution 0.1–0.14 at 540 nm	[12]
10	Rooster and chicken combs	Wash of the grounded raw material with ethanol with 1% of chloroform. Extraction with 3–3.5 volumes of water, acidified to pH 3–4 at 90–100°C during 40 – 60 min. Yield 0.09%.	Extracts are filtered, proteins removed at 60–80°C, 1–2 h with charcoal, then diethylaminoethyl-cellulose (1–1,5% from the extract volume); filtration at 30–40°C through polyvinyl-chloride membranes.	Hyaluronan ultrapure powder, protein content less than 0.1% (ovalbumin 0.001%)	[13]

(Continued)

Table 3.2 (Continued)

#	Source	Extraction	Purification	Final Product	Reference
11	Rooster combs	Before grinding the tissue is treated with ethanol in a ratio 1:2, then grind, treat with ultrasound (16–20 kHz 20–25 min), Extraction with water at 45–50°C 20–25 min. 55% of hyaluronic acid could be extracted.	Vacuum filtration of the extracts; HA 95% purity precipitation with ethanol at the ratio 1:3, drying.	Hyaluronan dry powder	[14]
12	Rooster combs or umbilical cord	Grinded raw material is frozen to (−20–70°C), 2 parts of water by weight added and the mixture is heated for 15–25 minutes at 95–100°C. Method increases the yield of HA in 3–4 times.	HA precipitation with acetic acid; ultrafiltration; lyophilization.	Hyaluronan dry powder	[15]
13	Rooster combs	The tissue is treated with ethanol in ratio 1:2, extracted with water with collagenase 0.03–0.04% to the tissue weight for 45–50 min, at 45–50°C, pH 6.8–7.0. As a result, increasing yield and better quality of HA.	Precipitation with ethanol at the ratio 1:3; vacuum filtration, vacuum drying, or sublimation.	Hyaluronan dry powder or solution	[16]
14	Rooster combs	Frozen tissue is treated with water at 55°C, grinded and adjust pH to 7.5. Proteinase is added and proteolysis is carried out for 3.5 h at 37°C. After filtration 5.6 g of the final product is obtained from 1 kg of the tissue.	Precipitation with cetylpyridinium chloride; the precipitated powder dissolved in 30% of ethanol with sodium chloride and re-precipitated with ethanol.	Hyaluronan dry powder	[17]
15	Rooster combs or umbilical cord	The combs are boiled in water for 45 min, grinded and heated for 4 h at 50°C and pH 7.5 with pronaze. Yield 6.7 g from 1 kg of the tissue.	Filtration; precipitation with cetylpyridinium chloride (CPC); The precipitant is dissolved in 30% ethanol with sodium chloride and treated with ammonium chloride in order to precipitate the final product.	Hyaluronan dry powder	[18]
16	Eggshell membrane	Wet eggshell membrane is treated with a yeast enzyme complex to reduce the eggshell membrane particles to a thick, almost clear slurry. The slurry is diafiltered using a 20 000 molecular weight cut-off membrane.	pH adjustment to 7.2; Cetylpyridinium chloride is added at 1:60 (v/v); centrifugation or filtration; ethanol added to the filtered HA solution at 2:1 ratio; centrifugation or filtration; the precipitate is dissolved in 0.2 M NaCl in 0.2 M phosphate buffer, pH 7.2; ethanol precipitation, filtration and acetone wash.	Hyaluronan dry powder	[19]

by water-chloroform extraction or by a mixture of chloroform-iso-amyl alcohol in a ratio of 5:1 [22]. To complete the removal of the proteins, after the precipitation HA is dissolved in water, the same volume of chloroform is added and the mixture is stirred intensively for 30–60 min until the emulsion is created. Then the mixture is left to create two layers – water and chloroform – with or without centrifuge. Usually the proteins are localized at the border situated between the water and chloroform phases.

Multiple repetitions of this procedure allow for the removal of a considerable amount of the protein impurities. However, HA of the animal origin contains covalently bonded peptides and proteins as well. In order to remove them, several proteolytic enzymes such as pepsin, trypsin, papain or pronase, could be used. These enzymes have different optimal activities depending on their temperature, pH, ionic composition and ionic media strength. In order to remove other mucopolysaccharides from the final product, a fractional precipitation with cetylpyridinium chloride [23], followed by dissolution of the complex in 0.2N sodium chloride could be done. Polysaccharides could be removed with ion-exchange chromatography, cellulose, gel-filtration on the agarose column or by sephadex. Other methods could also be used to purify hyaluronan, including ultrafiltration, sorption on the activated carbon, ion-exchange resin, electrodialysis, electrophoresis and ultracentrifusion with caesium chloride.

3.1.3 The Chemical Production of Hyaluronan from Chicken Combs

Three main methods have been used to chemically produce hyaluronan from chicken combs.

3.1.3.1 *Example 1*

The first method is a relatively simple and short process [5] that involves multiple water extractions of grounded rooster combs, precipitation from combined extracts of HA with acetic acid, treatment with base, enzymatic hydrolysis of the protein, ultrafiltration, ethanol precipitation and final dissolution of the precipitate hyaluronan in water.

First, the rooster combs are kept under cold water that must be periodically changed to maintain temperature. Then the combs are removed from the water, squeezed to drain and carefully grounded. Water is added to the grounded tissue at a comb/water ratio of 1:2.5 and brought to a boil, after which the extract is separated. The extraction process is repeated six times. It is interesting that the fifth and sixth extracts are free of lipids, the protein content is quite minimal and the HA content is comparable to the first four extracts. The fifth and sixth extracts are then combined and the polysaccharide is precipitated with acetic acid at pH 5 for 18.5 h. The formed precipitate is separated, dissolved in distilled water at a ratio of 1:3, and the pH is adjusted to 8 with sodium hydroxide solution.

Then a solution of papain (200 ml of 0.1% solution of papain per 1 kg of material) is added. Enzymatic hydrolysis of proteins is carried out for 20–40 h. Every 6–8 h the pH must be adjusted to 6–7 (to optimize enzyme activity). The HA mixture is purified by ultrafiltration and precipitation with 96% ethanol. The final product, sodium hyaluronate, is dried by lyophilization and the precipitate is then dissolved in a water-alcohol mixture containing 40% ethanol. The final purification of hyaluronan is performed by membrane filtration of a solution containing 30–50% ethanol.

3.1.3.2 Example 2

The second method of chemical production [9] includes double extraction of the grounded chicken comb with aqueous 5 to 25% solution of n-propyl or iso-propyl or *tert*-butyl alcohol with a combs/solvent ratio of 1:10–15. The extracts are combined and sodium chloride is added until two layers have been created. The aqueous layer is separated and HA is precipitated with ethanol. The precipitant is dried by lyophilization. The final product represents almost 50% of the initial content of HA in the combs with a protein content of less than 1%.

3.1.3.3 Example 3

The third method of the production of HA includes an initial wash of the tissue of the rooster combs with hot water at 60°C for 20–30 min. The two-stage extraction of the homogenized combs with physiological solution at 80–90°C takes place with a tissue/solvent ratio of 1:5. The extraction with the physiological solution excluded the contamination with nucleic acids (the lipids could be easily removed from the solution surface). The combined solutions are treated with 1–2% acetic acid. The obtained precipitant is dissolved in distilled water keeping the ratio precipitant/water 1:2, then 0.1% of sodium hydroxide added to the solution up to pH 7.0–7.3 and the solution heated slowly till 80°C and filtered. The filtered solution is lyophilized to afford sodium hyaluronate in the yield of 4% from the initial amount in the combs.

3.1.4 HA Production for Ophthalmology

An especially pure HA product for medical use, particularly for ophthalmology, was obtained in 1980 by Swedish company Pharmacia under the name 'Healon' [4]. The production method includes the following stages. The grounded animal tissue was treated with 95% ethanol denatured with chloroform for 24 h. Treatment was repeated several times until the solution stayed colourless and transparent. Then HA was extracted with mixture of water and chloroform 20:1. The mixture was stirred and let to stay without stirring for 24 h at 4–25°C, the mixture filtered, and the extraction repeated twice. Aqueous sodium chloride and chloroform 1:1 were added to combined extracts and a mixture was stirred for 3–5 h at 4–25°C. Then the mixture was kept until full fractions separated and organic fraction was isolated. The aqueous fraction was treated with hydrochloric acid till pH 4–5 and the equal volume of chloroform was added again. The procedure was repeated until the chloroform layer became transparent.

The alternative method to remove proteins is by treatment of the aqueous solution of hyaluronan with proteolytic enzymes. The proteolytic enzymes were added to the aqueous phase at pH 6–7 by addition of chloroform and stirring at 20–40°C for 5 days. The organic layer was separated, and the aqueous phase was sterilized by filtration trough Teflon filters and precipitated with three volumes of ethanol. Then the polysaccharide was again precipitated with ethanol and acetone and lastly washed three times with sterile acetone then dried in a vacuum.

The final product contained less than 0.5% proteins and peptides and had a molecular mass over 750 000 Da. The kinematic viscosity of 1% solution in the physiological buffer was 10^1–10^3 Stokes. The UV spectrum of 1.0% solution (E^{CM} 1%) at 257 and 280 nm (λмах): 257 (3), 280 (2). The typical yield is 0.8 g of hyaluronic acid from 1 kg of the chicken combs. Healon is sterile, non-pyrogenic and does not have antigen activity.

A highly pure product similar to Healon's with a molecular mass of 1 586 000, was described in another patent [24]. Using the described method, a tissue was treated with ethanol and then sodium hyaluronate was extracted with water. The proteins were removed with proteolytic enzymes, followed by chloroform extraction at pH 6–7. The sterilization was performed with cetylpyridinium chloride.

The patent [24] described the method of production of hyaluronan fractions with the average molecular mass from 250 000 to 350 000 Da. The proteins, which were still found in the extracts, were hydrolysed with papain, and then the resulting solution underwent ultrafiltration through a membrane. In addition to purification, ultrafiltration afforded separation from the fraction of 30 000 Da and lower, since such a low HA molecular mass could activate inflammation processes for when applied parenterally. The membrane can hold the HA fractions above 30 000.

The next ultrafiltration stage was carried out on another membrane that could hold molecules with a mass above 200 000 Da. The second fraction that passed through the membrane had the average mass of 50 000–100 000 Da. This product received the name 'Hyalastine'. The fraction that remained on the membrane had the average molecular mass of 500 000–730 000 and is called 'Hyalectin'.

S. Lorenzi and A. Romeo found an interesting method that prevents HA depolymerization during the process of the extraction [25]. It was achieved by addition of 1,10-phenantroline into an extraction solution. The purification from the proteins was done by proteases. A final purification stage was done through a transformation of HA into quarterly ammonium salt and the impurities were isolated by ion-exchange chromatography using dimethylsulfoxide or N-methylpyrrolidone as the solvents. The sterilization was performed with cetylpyridinium chloride in a phosphate buffer. The final molecular filtration allowed for the isolation of the fraction of the biopolymer with a molecular mass of 750 000–1 230 000 Da.

The product received the name 'HA-1'. Its protein content is not more than 0.2% (by J. Lowry [26], calculated for albumin), its static viscosity is in the range 14.5–21 dl/g (determined by a Ubbelohde type viscometer at 25°C at 0.15M), its solution of sodium chloride is at pH 7.0, and the UV absorption of 1.0% solution at 257 and 280 nm is not more than 1.0 AU). As determined on the instrument with inductive coupled plasma (ICP), the content of the sulfated mucopolysaccharides did not exceed 0.07% based on the sulfur content. The iron content did not exceed 10 ppm by the spectral atomic absorption or by ICP. The stability of the buffer isotonic solutions at pH 7.0 of HA-1, which underwent the normal ageing and thermal sterilization, was measured by the static viscosity and described as the reduction of the average molecular mass. The stability of HA-1 did not exceed the following limit data:

- 97% from the initial values at the storage for 6 months at 25°C
- 75% from the initial values at the sterilization (118°C) for 32 min
- 90% from the initial values at the sterilization (124°C) for 8 min.

The connective tissues and liquids of HA, as a rule, exist firstly as chondroitin sulfates in association with collagen [27,28] and other glycosaminoglycans. Purification methods must therefore include the purification of HA from these impurities. An extraction of hyaluronate with water and aqueous salt solutions is accompanied with the presence of a large amount of impurities of protein nature, other polysaccharides, nucleic acids, lipids and lipoproteins. The limitation of the protein removal fermentative methods depends on the activity of the proteolytic enzymes from the residual lipid content. The partial purification from the lipids is performed with hot water, acetone and ethanol. This method of base extraction requires the removal of the lipids and could lead to biopolymer destruction. As the precipitation reagents, cetyltrimethylammonium chloride and bromide are used more often [26,29]. The properties of the ammonium salts of the acidic glycosaminoglycans and proteins are quite different, which allows for them to be separated.

Using the quarterly ammonium salts, it is possible to perform the hyaluronan fractionation, which can be done by adding cetylpyridineammonium chloride to a solution of sodium hyaluronate in 0.25 N solution of sodium sulfate. The supernatant is diluted subsequently with aqueous sodium sulfate at 0.19, 0.174, 0.165, 0.130 and 0.06 N concentrations. After each dilution the solution is centrifuged and HA isolated from the precipitant by dialysis and re-precipitation with three volumes of ethanol. As a result, several fractions with different molecular masses are isolated. A similar result could be achieved by fractional precipitation with acetone or molecular ultra-filtration through the appropriate membranes [24,25].

Cetylpiridinum HA complexes are purified and fractionated by the differential re-precipitation and chromatography on sephadex and ion-exchange resins [25]. The proteins, peptides and nucleotides are removed from HA by the absorption on the activated carbon through ion-exchange resins like Dowex-1. The HA and mucopolysaccharides mixture is separated over the chromatographic columns with the sorbents made from the porous glass, sephadex, agarose, ion-exchange resins and other carriers and centrifugations in the density gradients. At the same time, physico-chemical parameters like the average molecular mass and molecular mass distribution of the biopolymer are analysed. Features of the other methods of HA isolation are presented in Table 2.2.

Material sources for the preparative isolation of HA usually include rooster combs, vitreous eye, umbilical cords, mammal skin and cartilage of sharks and whales. New methods of HA production are related to its biosynthesis in cell culture and creation of the gene-engineered microorganism shams.

3.2   Bacterial Methods of Hyaluronic Acid Production

Modern HA production is done by microbiological methods; extensive amount of the scientific and patent literature about the production of hyaluronic acid by microbiological method has been accumulated and is available for reference [30].

A limited amount of gram-positive bacteria (*Streptococcus* sp. and *Pasteurella* sp.) are able to synthesize a polysaccharide from the upper capsule (1 mm thick) they create [31,32]. The majority of such microorganisms are pathogenic to humans and animals, but they are also able to parasite in the intercellular space of the mammal's tissue. That is why there is high demand for these microorganisms:

1. They should not be pathogenic to humans and show no haemolytic activity.
2. They should be able to synthesize high molecular weight HA at high speed.
3. They should not show any hyaluronidase activity so as not to hydrolyse the high-molecular product.

4. They should be stable when stored.
5. They should use a substrate as fully as possible.

At the present time such non-haemolytic and hyaluronidase-negative strains are usually obtained by non-direct chemical and UF-induced mutagenesis, followed by selection or cloning of the hyaluronidase gene in order to transform *Streptococcus* sp. into non-pathogenic bacteria [33]. Of all the hyaluronan producers, the most appropriate strain is streptococcus of the group C, that is *Streptococcus equi* because it produces the highest yield of HA [30].

Biotechnological methods of hyaluronan production from bacterial strains involve cultivation in selected conditions where the polysaccharide capsule is formed during the stage of logarithmic growth on the surface of the bacterial cells. But at the stationary growing stage, HA can move into the cultural liquid and a capsule could become thin or disappears completely [30]. At the end of the process, up to 1–6 g of the desirable product could accumulate in 1 l of the cultural liquid. HA accumulation could be controlled by the measurements of the viscosity of the cultural liquid.

The production of HA from the cultural media involves the removal of the microorganisms, removal of low molecular weight substances by ultrafiltration, precipitation with organic solvent and purification of the final product using the methods described earlier for the animal tissues (Table 2.2). At the end, the commercial product appears as white powder, flakes or a 1% solution (the solution has an average molecular weight of 1 000 000 Da and can have varying molecular weight distributions).

3.3   Hyaluronan Destruction during Production, Storage and Sterilization

One of the most important tasks during the extraction of polysaccharides from animal tissues and microorganisms is to maintain the integrity of the macromolecule. Since HA has a high molecular weight (up to 9 000 000 Da) and straight linear structures up to 2.4 mm [34,35], it is very sensitive to mechanical, thermal, chemical and enzymatic perturbations that could result in a decrease of molecular weight or degree of polymerization during the processes of extraction, purification, sterilization and storage.

Long polysaccharide chains of HA in aqueous solutions could fragment under various physical and chemical conditions including ultrasound, radiation, high temperature, extreme pH, exposure to oxidation reagents, free radicals and dynamic motion. However, it must be primarily considered that HA could be affected by the hyaluronidases, which would result in enzymatic degradation. As a result, usually extraction and purification provides the commercial products of HA with an average molecular mass of $1-2 \times 10^6$ Da and quite a wide mass distribution (Figure 3.1).

It is important to control the molecular weight (or the length) of the polysaccharide during production because it has a direct effect on the biological and therapeutic properties of HA. For example, oligosaccharides with a molecular weight of about 10 000 Da could induce angiogenesis [36] and tetra- and hexasaccharides could activate dendritic cells [37]. (The functions of oligosaccharides are described in Chapters 2 and 6.) Thus, it is important for medical and cosmetic applications of HA to obtain the oligosaccharides with the required molecular weight.

3.4   Enzymatic Destruction of Hyaluronan

The destruction of HA initially takes place by hyaluronidase of the animal tissues or bacterial hyaluronidase. In animal tissues the depolymerization of the biopolymer is done by hyaluronidases (endoglycolases), beta-glucuronidase and beta-N-acetylhexozaminidase (exoglycolase). Hyaluronidase's actions lead to hydrolysis of the bonds along the polysaccharide chain which attribute to the reduction of the molecular weight of the polysaccharide during extraction. Other enzymes affect the chain terminus and decompose the polymer at the end of the molecule. The reaction rate depends on the length of the chain – the longer the chain, the higher the decomposition rate until it results in oligosaccharides and further decomposes to the mixtures of octa-, hexa-, tetra- and disaccharides. With an increase in the degree of biopolymer depolymerization, the activity of endohyaluronidases reduces but the activity of exoglycolases increases. Also, exoglycolases transform even oligosaccharides into odd ones (e.g., tri- and penta-saccharides) and increase the concentration of free glucuronic acid and acetylglucosamine [2]. In order to supress the activity of the enzymes and non-controlled enzymatic hydrolysis, the extraction of HA is recommended to carry out at low temperature and, in the presence of chloroform [38–40], treat the combs with hot water, carry out extraction at high temperature (60–100°C) [5,7,10,12] and use a low pH up to 3–4 [7]. Many activators and inhibitors of animal tissue hyaluronidase are well known (see Chapter 2). The usage of hyaluronidase inhibitors allows for the rate of depolymerization of hyaluronan to be reduced during its extraction [25]. The reduction of the rate of the enzymatic destruction during microbiological synthesis could be achieved by the use of hyaluronidase-negative bacterial strains.

Next is the detailed description of the main characteristics of the hyaluronidase enzymes.

3.4.1   Hyaluronidase Classification

Recently, numerous studies regarding the localization, structure and properties of hyaluronan have been published [41]. However, the systematization initially proposed by K. Meyer is still relevant. Based on Meyer's classification [42], hyaluronidases could be divided into three types based on source enzyme criteria, the type of the reaction that the enzyme could catalyse and the final products.

Type 1. Hyaluronidases of the testicular type (hyaluronate-endo-β-N-acetylhexozaminidazes)
- *Type 1a.* Testicular hyaluronidase, which is contained in the animal seminal glands and sperm;
- *Type 1b.* Lizosomal hyaluronidase, which is present in the lyzosomes of the different cells, blood serum, synovial liquid;
- *Type 1c.* Submandibular hyaluronidase, which is present in animal saliva and salival glands.

Type 2. Hyaluronidase from leech saliva. The final products of hydrolysis by both Type 1 and Type 2 hyaluronidases are tetrasaccharides, which possess amino sugar at their reducing terminus and are activated with the enzymes of the Type 1. Additionally, the testicular and lyzosomal hyaluronidases are capable of trans-glycosylase activity and can transfer disaccharide fragments between the molecules of the substrate. The testicular hyaluronidase is active within the pH range of 4.0–7.0, but the lyzosomal and submandibular hyaluronidases are only active in the narrower range of pH 3.5–4.5. At pH 8.2 the enzymes are not active.

Type 3. Microbial hyaluronidases (e.g. *Streptococcus* hyaluronidase). Microbial hyaluronidases hydrolyse β-N-acetylaminoglycoside bonds of a substrate and simultaneously dehydrate the residue of uronic acid at the non-reducing terminus of the molecule. Substrate specificity of bacterial hyaluronate lyases varies considerably in the different species of microbe producers. Hyaluronidase of *Streptococcus pneumoniae* has the highest substrate specificity; it hydrolyses HA alone and does not destroy other glucosaminoglycans [43]. The hyaluronate lyase, when isolated from *Streptococcus pneumoniae*, reaches optimal activity at pH 6.0 with the Michaelis constant with respect to HA being equal to 3.8×10^{-4} mol/l (in terms of Michaelis–Menten kinetics) [44]. The presence of Ca^{2+} (about 10 mM) is necessary in order to show the maximum enzyme activity.

3.4.2 Properties and Functions of Hyaluronidases

Mammal hyaluronidases such as testicular hyaluronidase hydrolyse β-1,4-glycoside bonds inside the HA chain, forming oligosaccharides of different sizes. After prolonged incubation, the final products are mainly hexa- and tetrasaccharides [37]. Mammal hyaluronidases do not have absolute substrate specificity and could hydrolyse chondroitin-4-sulfate, chondroitin-6-sulfate and dermatansulfate. In addition, these enzymes have transglycosilate activity (more details about hyaluronidase isoforms and properties in the human somatic tissues and body fluids are described in [45]).

In microbiologic technologies of HA production it is fragmented by the bacterial hyaluronidases that are transported into the cultural media. They are called hyaluronate lyases because they are endo-β-acetylhexosamine eliminases. The most typical representative of this class of enzymes are the *Streptococcus* hyaluronate lyases. The bacterial hyaluronate lyases hydrolyse endo-β-1,4-glycoside bonds in HA via the β-elimination reaction, the result of it is the formation of 4,5-unsaturated oligosaccharides of different lengths.

The final product of the enzymatic cleavage is a 4,5-unsaturated disaccharide. Therefore, depolymerization of HA with bacterial hyaluronate lyases can be controlled not only by viscometric but also spectrophotometric methods as well. Unlike bacterial hyaluronidase, those from animal tissues are able to fragmentize HA, which could result in the production of saturated oligosaccharides.

The most appropriate enzyme for the fragmentation of pure HA is testicular hyaluronidase. It hydrolyses HA through the formation of the saturated oligosaccharides (which are important for therapeutic use), is active in the wide pH range 4.0–7.0 and has high thermal stability (it can maintain enzymatic activity up to 50°C).

Bacterial hyaluronate lyases have a narrow substrate specificity. They split HA that has a higher rate in order to lower the molecular weight of the final products. At the same time, they create unsaturated oligosaccharides that are likely to have different therapeutic properties to saturated oligosaccharides. Several unsaturated oligosaccharides were isolated after a treatment with hyaluronate lyases in order to study their properties [46]. The bacterial hyaluronate lyases could help to study kinetics of HA depolymerization using spectrophotometric methods that allow for the reaction products' level of unsaturation to be identified.

Further details on the effect of HA enzyme concentration and ionic strength of the media on the kinetics of the hydrolysis are compiled in the references [47].

After the enzyme treatment, the level of fragmentation of HA and the average molecular weight of the oligosaccharides can be determined by viscometry. The analysis of the molecular weight distribution and isolation of the several fractions could be performed by gel-chromatography using the columns with sepharose 4B (Figure 3.1) or ion-exchange chromatography [46].

HA could also be hydrolysed with dimethylsulfoxide/hydrochloric acid [46] or with diluted hydrochloric acid [48], followed by isolation of the oligosaccharide fractions. However, the enzymatic fermentation is more economically viable. An example of the typical production of the low molecular weight of HA oligosaccharides is described in [37], wherein HA was fragmented with testicular hyaluronidase for 12 h in 1M sodium-acetated buffer at pH 5.0 and 37°C. The fragments were isolated on the column Biogel P10. The size of the fragments was determined by electrophoresis in 30 polyacrylamide gel.

3.5 Non-Enzymatic Destruction of Hyaluronan

In addition to enzymatic hydrolysis, HA is quite sensitive to different non-enzymatic treatments; that is acid-alkaline hydrolysis, oxidation-reduction and others.

3.5.1   Acid-Base Hydrolysis of Hyaluronan

First of all, HA is sensitive to acid-alkaline hydrolysis. Even a slight HA acidification by an acetic acid solution leads to an irreversible decrease in viscosity by 2.5 times [34]. Mineral acids can completely hydrolyse HA into glucuronic acid, glucosamine, acetic acid and carbon dioxide. And, in a short period of time, diluted sulfuric acid hydrolyses hyaluronic acid to form disaccharide crystals [35].

3.5.2   Oxidation-Reduction Depolymerization of Hyaluronan

Through the treatment process of redox media, degradation of polysaccharide macromolecules is done through a free radical mechanism whereby free radicals are formed in the presence of ascorbic acid, oxidase and oxygen [49]. HA could be depolymerized by the ions Fe^{2+} and Fe^{3+} in the presence of reducing reagents, particularly ascorbic acid. This explains why HA isolated in the atmosphere of nitrogen or argon has a higher polymerization level compared to HA isolated in the air [1].

During lyophilization of hyaluronan solutions, polysaccharide degradation is initiated with phosphate ions [50]. For purification from proteins, proteolitic enzymes could also be used to reduce the viscosity of the biopolymer, particularly papain SH-groups that are reducing agents and accelerate the decomposition of hyaluronan [34]. Trypsin, which contains Fe^{3+} ions, could also be used for the purification of hyaluronan and to initiate the depolymerization process. The use of 8-hydroxyquinoline prevents HA viscosity reduction [25]. In order to maintain the polymerization level of hyaluronan, the initial tissue must be thoroughly washed from blood, which contains ions of iron, copper and phosphate.

For medical use, the hyaluronan solutions must be sterilized. Sterilization is usually achieved by autoclaving at 120–130 °C or by ionizing gamma radiation. In both cases there is a significant depolymerization of the biopolymer and loss of its initial therapeutic activity. There are methods to protect the hyaluronan solutions against depolymerization by adding amino acids, boric acid and glycerol, hydroproline sulfate, uric acid and phenolic compounds, for example pyrogallol [40].

The most efficient method is described in the patent [40]. 1g of HA was dissolved in 100 ml of 0.9% sodium chloride solution. The resulting solution was purged with argon for 30 min at 10 ml/min and 0.02–0.04 g of copper sulfate and 0.06–0.08 g of sodium ascorbate added. The solution was then poured into 1 ml ampoules under argon, and the ampules sealed and sterilized by a 15 kGy dose of gamma radiation. This method allows for the prevention of hyaluronan destruction in solution during irradiation and prolonged storage.

3.6   Quality of Hyaluronan Commercial Products of Animal and Bacterial Origin

The commercial product of hyaluronan must correspond to several quality requirements by meeting the acceptance criteria. In Table 3.3 is the typical product specification with several quality parameters and acceptance criteria.

Additional characteristics could include average molecular weight, nitrogen, sulfur content and other elements.

Several methods to determine the hyaluronan content have been described. According to the aforementioned Dische method, glucuronic acid content could be found in HA [51]. The Morgan–Elson method, modified by Bitter, could be used for evaluation of HA content by N-Acetylaminoglucose content [52]. If the product does not contain other polysaccharides as impurities, the content of glucuronic acid and N-acetylaminoglucose should be close to 1:1.

The absence or low level of sulfur content in a hyaluronan product is another criterion that suggests an absence or small amount of sulfated mucopolysaccharides. The presence of other impurities could be qualitatively determined by absorption of the biopolymer on the filter paper and colouring by toluidine blue [53]. Hyaluronan gives a strong bright blue colour, whereas chitosan and other polysaccharides that contain amino groups are revealed as weak spots. Nucleic acids (DNA and RNA in the concentration of 10 mg/ml and more) appear as a weak blue spot with a weaker ring around the spot. The neutral polysaccharides, proteins and small molecules do not appear on the paper chromatography, though many acidic polysaccharides do. The number of nucleic acid impurities could be determined more precisely by chemical methods [4], or spectrophotometrically by hyperchromic effect [54]. The protein concentration is usually determined by the Lowry method [26] using the Folin–Chokalteu reagent. This is the most sensitive method of quantitatively determining proteins since it is based on a combination of biuret reaction and the reduction of the Folin–Chokalteu reagent by tyrosine and tryptophan residues in the protein.

It should be noted, however, that the sensitivity of the method is reduced because HA in animal tissues is connected mainly with fibrous proteins of the intercellular matrix (collagen, elastin, etc.), which contain very small amount of these amino acids. For the same reason, spectrophotometric methods for determining of the protein concentration at 280 nm

Table 3.3   *Hyaluronan commercial product typical specification*

No.	Quality parameter (test)		Acceptance criteria
1	Appearance		In organic solvent has a fibrous structure in the form of individual fibres and the gloms; in dried form is a powdery substance
2	Colour		White material, could be slightly creamy
3	Odour		Weak odour, characteristic of certain type of raw material
4	Mass content in dry compound, % not more than	- moisture	6.0
		- proteins	0.5
		- ash	4.0
5	Kinematic viscosity, m²/c		2.0×10^{-4}
6	pH 1% solution		6.5 ± 1.0
7	Heavy metal salts, mg/kg, not more than	- lead	5.0
		- arsenic	5.0
		- mercury	1.0
8	Toxicological data	- acute toxicity LD_{50} on skin	More than 2500
		- irritating action on skin one time action	Not present
		- irritating action on skin multiple actions	Not present
		- sensibilizing action	Not present
		- irritating action to the rabbit eye mucous membrane	Not present
9	Microbiological data	- Microbes in 1g of the product	
		- *Enterobacteriaceae*	Not more than 100
		- *Pseudomonas aeruginosa*	Not present
		- *Staphilococcus aereus*	Not present
		- Fungus *P. candida*	Not present
		- Mildew on 1g of the product	Not present

(the maximum absorption of aromatic amino acids – tryptophan, tyrosine and phenylalanine) have a reduced sensitivity as well. Therefore, a more accurate and reliable method to determine the concentration of proteins in hyaluronan is the method by which the kinetics of the concentration of amino acids increase in the proteolysis conditions (e.g. pronase) is measured. The growth dynamic of the free amino acid concentration is determined by the standard reaction with ninhydrin [55].

A total protein content of 0.5% and less in a commercial product is considered safe regarding the product's immunogenic properties. Among the protein's impurities, the most antigenetic properties are attributed to ovalbumin, which has the ability to penetrate through skin and mucous membranes by bypassing the immune barrier. The ovalbumin content could be determined by solid phase immune-enzyme test-system with sensitivity of −0.5 ng/ml. Its content must be lower than 0.2% from the overall protein content.

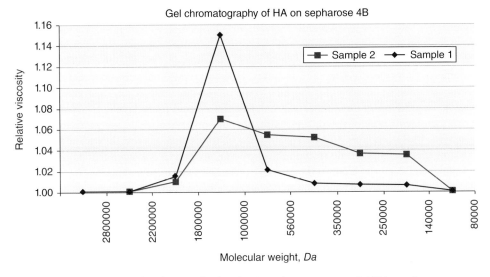

Figure 3.1   *Molecular weight distribution of two commercial HA products*

The value of the average molecular weight of hyaluronan required by specification does not reveal its molecular weight distribution. Such data could be evaluated by gel-chromatography using sepharose media. Figure 3.1 shows a molecular weight distribution spectrum of two different commercial HA products of the bacterial origin [56,57].

These data, produced by two different companies through microbiological fermentation, show the wide range of molecular weight of HA. However, both certificates of analysis state the same average molecular weight – 1 000 000 Da. Based on our experience, commercial products from different suppliers could vary in their molecular weight distribution depending on the batch of the commercial product, even from the same supplier. This important characteristic depends on many different factors such as source and extraction method, purification, sterilization and storage.

The evolutionary conservatism of hyaluronan is well-known fact. This means that during evolution of living organisms, the chemical structure of HA does not change. Evolutionary conservatism is characteristic of vital biological structures. Because of this trait, any HA extracted from different sources are compatible with the cells and tissues of human body; do not cause allergies, pyrogenic reactions, irritation or other negative application consequences. However, the toxicological analyses of commercial HA products include only a limited number of quantitative assessments of microbiological contaminants and reactions on animal skin after the application of HA (Table 3.2).

At the present time, the market share of commercial hyaluronan of bacterial origin is significantly higher than the products extracted from animal sources, the reason being that microbiological synthesis affords a significantly higher yield: the fermentation of the strain *Streptococcus* sp. resulted in a polysaccharide concentration of 5–6 g per 1 l of culture medium. Comparatively, the yield of HA in the extraction from the chicken combs is from 0.5 g to 6 g per 1 kg of raw material. Because the purification of bacterial HA from proteins and peptides is easier, so it is therefore easier to obtain the HA product with a molecular mass of ~1 million Da and protein content of less than 0.05%. The HA products from animal tissues has an average molecular weight of 750 000 Da and a protein content of about 0.5% and possess covalently bonded low molecular weight proteins. Purification from these proteins requires strict manipulations that reduce the molecular weight of hyaluronan. Because contaminating proteins are major allergens, the degree of protein purification is a main characteristic of drug safety. However, one should consider that the degree of immunogenicity of protein impurities in the product of bacterial origin could be higher than in the products of the animal origin despite the overall lower levels.

As mentioned previously, during extraction the polysaccharides are partially decomposed by hyaluronidases to become oligosaccharides that possess hexosamine on the reducing end and glucuronic acid on the non-reducing end. It is known that animal hyaluronidases and bacterial hyaluronidases have different mechanisms of action. Each of them cleave β-(1-4)-glucoside bonds in the biopolymer, but contrary to animal hyaluronidases, the bacterial hyaluronitliases perform intermolecular hydrolysis with removal of water. As a result, different disaccharides with the double bond between C4 and C5 carbon atoms of the glucuronic acid are created [58]. The presence of the double bonds helps to evaluate the level and kinetics of the enzymatic depolymerization by measuring the double bond accumulation by spectrophotometric analysis at 232 nm.

The presence of nucleic acids in HA products isn't usually evaluated nor stated in the certificate of analyses. Although the antigenicity of RNA and DNA impurities is low, due to presence of the residual proteins they may be partially responsible for total antigenicity. The material from animal tissue could contain sulfated polysaccharides as a contamination. There are methods to separate the polysaccharides that can reduce the concentration of these impurities to 0.07% based on the sulfur content [25]. In addition, the sulfated mucopolysaccharides, like hyaluronan, are natural polysaccharides of animal tissues and thus cause no inflammatory reactions.

The real problem during the enzymatic synthesis of hyaluronan could be the cultivation of pathogens in the reactors. The creation of genetically modified microorganism strains that synthesize the polysaccharide is a potential danger in the transfer of transgenic DNA (RNA) to the other strains and parasitic systems [59]. Considering the fact that most of the organisms that synthesize HA are, to a certain degree, pathogens for advanced animals and humans and parasites in the intercellular matrix of mammalian tissues, from the toxicological point of view HA products from animal tissues may be more desirable for medical applications. However, this issue requires further in-depth study.

References

[1] Ignatova, E.Y., Gurov, A.N. (1990) Principles of extraction and purification of hyaluronic acid. (In Russian). *Chemical-Pharmaceutical Journal*, **24** (3), 42–46. Игнатова Е. Ю., Гуров А.Н. (1990) Принципы извлечения и очистки гиалуроновой кислоты *Химико-фармацевтический журнал*, **24** (3), 42–46.

[2] Radaeva, J.F., Kostin, G.A., Zinevsky, A.V. (1997) Hyaluronic acid: biological role, structure, synthesis, isolation, purification and application. (In Russian). *Applied Biochemistry and Microbiology*. **33** (2), 133–137. Радаева, И.Ф., Костина, Г.А., Зиневский А.В. (1997) Гиалуроновая кислота: биологическая роль, строение, синтез, выделение, очистка, применение. *Прикладная биохимия и микробиология*. **33** (2), 133–137.

[3] Swann, D.A. (1968) Studies on hyaluronic acid. The preparation and properties of rooster comb hyaluronic acid. *Biochimica et Biophysica Acta*, **156**, 17–30.

[4] Balazs, E.A. (1977) Ultrapure hyaluronic acid and the use thereof. US Patent 4.141.973, filed Oct. 25, 1977, and issued Feb. 27, 1979.

[5] Fedorischev, I.A., Chernyshov, A.A., Epiphanov, A.E. (1999) The method of obtaining hyaluronic acid (in Russian). Russian Federation Patent RU 2157381, filed Mar. 01, 1999 and issued Oct. 10, 2000. Федорищев, И. А., Чернышов, А.А., Епифанов, А. Е. Способ получения гиалуроновой кислоты. Пат. Ru № 2,157,381.// БИ 2000. 10.10.

[6] Tulupova, G.B., Muchtarov, E.I., Plechanova, N.Y. et al. (1989) The method of obtaining hyaluronic acid (in Russian). Soviet Union Patent SU 1616926, filed on May 05, 1989, and issued Nov. 07, 1990 Стекольников, Л.И., Рыльцев, В.В., Игнатюк, Т. Е. Способ получения гиалуроновой кислоты. АС № 1616926 СССР // БИ 1990.- № 48.-С. 85.

[7] Stekol'nilov, L.I., Kornilova, A.A. (1993) The method of obtaining hyaluronic acid (in Russian). Russian Federation Patent RU 2074863, filed Aug. 23, 1993, issued Mar. 10, 1997. Стекольников, Л.И., Корнилова, А.А. Способ получения гиалуроновой кислоты. Патент РФ № 2074863.// Б. И. 1997. 03. 10.

[8] Stacey, M., Barker, S.A. (1962) *Carbohydrates of Living Tissues.* Van Nostrand, London. ССтейсн. М., Баркер, С. (1965) *Углеводы живых тканей.* Мир, Москва.

[9] Ryashentsev, Y.V., Nikolski, S.R., Vainerman, E.S. et al. (1991) The method of obtaining hyaluronic acid. (In Russian). Russian Federation Patent RU 2017751, filed May 22, 1991, issued Aug. 15, 1994. Ряшенцев, Ю.В., Никольский, С.Р., Вайнермен, Е.С. Способ получения гиалуроновой кислоты. Патент № 2017751 РФ.// Б.И. -1994.-№ 5.- С. 75–76.

[10] Radaeva, I.F., Kostina, G.A. (1995) The method of obtaining hyaluronic acid (in Russian). Russian Federation Patent RU 2102400, filed Aug. 08, 1995, issued Jan. 20, 1998. Радаева, И.Ф., Костина, Г.А. Способ получения гиалуроновой кислоты. Пат РФ № 2102400.// Б.И. 1998. 01.20.

[11] Stekol'nikov, L.I., Ryltsev, V.V., Virnik, R.B. et al. (1992) The method of obtaining hyaluronic acid (in Russian). Russian Federation Patent RU 2046801, filed Feb. 11, 1992, issued Oct. 27, 1995. Стекольников, Л. И., Рыльцев, В.В., Вирник, Р.Б. и др. Способ получения гиалуроновой кислоты. Пат. РФ № 2046801.// БИ.- 1995. 10.27.

[12] Muchtarov, E.I., Tulupova, G.B., Gromov, I.Y. (1992) The method of obtaining hyaluronic acid preparations. (In Russian). Russian Federation Patent RU 2055079, filed Feb. 04, 1992, issued Feb. 27, 1996. Мухтаров, Э.И., Тулупова, Г.Б., Громов, И.Ю. Способ получения препарата гиалуроновой кислоты. Пат. РФ № 2055079.// Б.И. 1996. 02.27.

[13] Stekol'nikov, L.I., Samoilenko, I.I., Kornilova, A.A. (1993) The method of obtaining hyaluronic acid (in Russian). Russian Federation Patent RU 2074196, filed Aug. 23, 1993, issued Feb. 27, 1997. Стекольников, Л.И., Самойленко, И.И., Корнилова, А.А. Способ получения гиалуроновой кислоты. Пат. РФ № 2074196.// Б.И. 1997. 27.02.

[14] Antipova, L.V., Polenskich, S.V., Aleksyuk, M.P. (1996) The method of obtaining hyaluronic acid. (In Russian). Russian Federation Patent RU 2114862, filed May 31, 1996, issued Jul. 10, 1998. Антипова, Л.В., Поленских, С.В., Алексюк, М.П. Способ получения гиалуроновой кислоты. Пат РФ № 2114862.// Б.И. 1998. 07.10.

[15] Samoilenko, I.I., Epiphanov, A.E. (1997) The method of obtaining hyaluronic acid (in Russian). Russian Federation Patent RU 2115662, filed Jul. 21, 1997, issued Jul. 20, 1998. Самойленко, И.И., Епифанов, А.Е. Способ получения гиалуроновой кислоты. Пат РФ № 2115662.// Б.И. 1998. 07.20.

[16] Antipova, L.V., Polenskich, S.V. (2001) The method of obtaining hyaluronic acid (in Russian). Russian Federation Patent RU 2186786, filed Mar. 26, 2001, issued Aug. 10, 2002. Антипова, Л.В., Поленских, С.В. Способ получения гиалуроновой кислоты. Пат. РФ № 2186786.// Б.И. 2002. 08.10.

[17] Miyazaki, T., Okuyama, T. (1977) Isolation of acidic polysaccharide. Japanese patent JPS52105199 (A), filed Feb. 15, 1976 and issued Sep. 03, 1977.

[18] Miyazaki, T., Tanaka, S., Takahashi, S. (1977) Extraction of acidic polysaccharides. Japanese patent JPS52145594 (A), filed Jan. 18, 1976, and issued Dec. 03, 1977.

[19] Long, F.D., Adams, R.G., DeVore, D.P. (2005) Preparation of hyaluronic acid from eggshell membrane. US Patent 6946551 (B2), filed Jul. 9, 2003, issued Sep. 20, 2005.

[20] Cleland, R.L., Wang, J.L. (1970) Ionic polysaccharides. Dilute solution properties of hyaluronic acid fractions. *Biopolymers,* **9** (3), 799–810.

[21] Bychkov, S.M., Kolesnikov, M.F. (1964) The method of obtaining hyaluronic acid (in Russian). Soviet Union Patent SU 219752, filed on Apr. 04, 1964, and issued May 17, 1968. Бычков, С.М., Колесников, М.Ф. Способ получения гиалуроновой кислоты. А.с.№ 219752, СССР.// Б.И. 1968.- № 19.-С. 90.

[22] Akasaka, H., Arai, T., Komasaki H. (1985) Fermentation method for producing hyaluronic acid. US Patent 4801539, filed May 9, 1985, and issued Jan. 31 1989.

[23] Laurent, T.C. (1970) Structure of hyaluronic acid, in *Chemistry and Molecular Biology of the Intercellular Matrix.* (ed. Balazs E.A.) London and New York. Academic Press pp. 703–732.

[24] Della, V.F., Romeo, A., Lorenzi, S. (1984) Hyaluronic acid fractions having pharmaceutical activity, methods for preparation thereof, and pharmaceutical compositions containing the same. European Patent EP0138572, filed on Oct. 10, 1984, and issued Jul. 25, 1990.

[25] Lorenzi, S., Romeo, A. (1992) Procedure for the purification of hyaluronic acid and fraction of pure hyaluronic acid for ophthalmic use. European Patent EP0535200 B1, filed on Apr. 16, 1992, and issued Nov. 3, 1999.

[26] Lowry, O.H., Rosebrough, N.J., Farr, L.A., Randall, R.J. (1951) Protein measurement with the folin phenol reagent. *Journal of Biological Chemistry,* **193**, 265–275.

[27] Brimacombe, J.S., Webber, J.M. (1964) *Mucopolysaccharides: Chemical Structure, Distribution and Isolation.* Elsevier Publishing Co, Amsterdam.

[28] Stepanenko, B.N. (1977) *Chemistry and Biochemistry of Carbohydrates (Polysaccharides).* (In Russian). Vyschaya shkola, Moscow. Степаненко Б. Н. *Химия и биохимия углеводов (полисахариды).* М.: Высшая школа.- 1977.-С. 285.

[29] Scott, J.E., Cummings, C., Brass, A., Chen, Y. (1991) Secondary and tertiary structures of hyaluronan in aqueous solutions. *Biochemical Journal,* **274**, 699–703.

[30] Beloded, A.B. (2008) *Microbiological synthesis and degradation of hyaluronic acid by Streptococcus species bacteria* (in Russian). PhD Thesis. Moscow State University. Белодед, А.В. *Микробиологический синтез й деградация гиалуроновой кислоты бактериями* р. Дисс. На соиск. уч.степ. к. б. н. М. 2008.-С. 187.

[31] Chong, B.F., Blank, L.M., Mclaughlin, R., Nielson L.K. (2005) Microbiological hyaluronic acid production. *Applied Microbiology and Biotechnology,* **66**, 341–351.

[32] DeAngelis, P.L. (1996) Enzymological characterization of the *Pasteurella multocida* hyaluronic acid synthase. *Biochemistry,* **35**, 9768–9771.

[33] Winder, B., Behr, R., Von Dollen S. et al. (2005) Hyaluronic acid production in *Bacillus subtilis. Applied and Environmental Microbiology,* **71**, 3747–3752.

[34] Balazs, E.A. (ed.) (1970) *Chemistry and Molecular Biology of the Intercellular Matrix.* Academic Press, New York.

[35] Shimada, E., Matsumura, G.J. (1977) Molecular weight of hyaluronic acid from rabbit skin. *Journal of Biochemistry.-* **81** (1), 79–91.

[36] West, D.C., Hampson, I.N., Arnod F., Kumar S. (1985) Angiogenesis induced by degradation products of hyaluronic acid. *Science,* **228**, 1324–1326.

[37] Termeer, C.C., Hennies, J., Voith, U., et al. (2000) Oligosaccharides of hyaluronan are potent activators of dendritic cells. *Journal of Immunology,* **165**, 1863–1870.

[38] Boas, N.F. (1949) Isolation of hyaluronic acid from the cocks comb. *Journal of Biological Chemistry,* **181**, 573–575.

[39] Balazs, E.A., Jeanloe, R.W. (eds) (1966) *The Amino Sugars: The Chemistry and Biology of Compounds Containing Amino Sugars II B. Metabolism and Interactions.* Academic Press, New York.

[40] Samoilenko, I.I., Stekol'nikov, L.I. (1992) The method of stabilization of hyaluronic acid solutions (in Russian). Russian Federation Patent RU 2051154, filed Nov. 02, 1992, issued Dec. 27, 1995. Самойленко И. Н. Стекольников Л. И. Способ стабилизации растворов гиалуроновой кислоты.// Пат. РФ № 2051154. Бюл. 1995. 11.07.

[41] Frost, G.I., Csóka, T., Stern R. (1996) The hyaluronidases: a chemical, biological and clinical overview. *Trends in Glycoscience and Glycotechnology,* **8**, 419–434.

[42] Meyer, K. (1971) Hyaluronidases, in *The Enzymes,* vol. **5**, (ed. P.D. Boyer), Academic Press, New York, pp. 307–320.

[43] Li, S., Kelly, S.J., Lamani, E., et al. (2000) Structural basis of hyaluronan degradation by Streptococcus pneumoniae hyaluronate lyase. *EMBO Journal,* **19**, 1228–1240.

[44] Kelly, S.J., Taylor, K.B., Li, S., Jedrzejas, M.J. (2001) Kinetic properties of Streptococcus pneumoniae hyaluronate lyase. *Glycobiology,* **11**, 297–304.

[45] Fiszer-Szafarz, B., Litynska, A., Zou, L. (2000) Human hyaluronidases: electrophoretic multiple forms in somatic tissues and body fluids: Evidence for conserved hyaluronidase potential N-glycosylation sites in different mammalian species. *Journal of Biochemical and Biophysical Methods,* **45**, 103–116.

[46] Xu, H., Ito, T., Tawada, A. (2002) Effect of hyaluronan oligosaccharides on expression of heat shock protein 72. *Journal of Biological Chemistry,* **277**, 17308–17314.

[47] Astériou, T., Deschrevel, B., Gouley F., Vincent, J.C. (2002) Influence of substrate and enzyme concentration on hyaluronan hydrolysis kinetics catalyzed by hyaluronidase, in *Hyaluronan: Proceedings of an International Meeting, September 2000, North East Wales Institute, UK* (eds J.F. Kennedy, G.O. Phillips, P.A. Williams, V.C. Hascall), Woodhead Publishing Ltd, Cambridge, pp. 249–252.

[48] Tommeraas, K., Melander, C. (2008) Kinetics of hyaluronan hydrolysis in acidic solution at various pH values. *Biomacromolecules,* **9**, 1535–1540.

[49] Swann, D.A. (1967) The degradation of hyaluronic acid by ascorbic acid. *Biochemical Journal,* **102**, 42C–44C.

[50] Kuo, J.W. (2006) *Practical Aspects of Hyaluronan Based Medical Products.* Taylor & Francis, Boca Raton, p. 37.

[51] Dishe, Z. (1947) A new specific color reaction of hexuronic acid. *Journal of Biological Chemistry,* **167**, 189–198.

[52] Bitter, T., Muir, H.M. (1962) A modified uronic acid carbazole reaction. *Analytical Biochemistry,* **4**, 330–334.

[53] Blumenkrantz, N. (1957) Microtest for mucopolysaccharides by means of toluidine blues with special reference to hyaluronic acid. *Clinical Chemistry,* **3**, 696–702.

[54] Voet, D., Gratzer, W.B., Cox, R.A., Doty, P. (1963) Absorption spectra of nucleotides, polynucleotides, and nucleic acids in the far ultraviolet. *Biopolymers,* **1**, 193–208.

[55] Harding, V.J., Warneford, F.H.S. (1916) The ninhydrin reaction with amino-acids and ammonium salts. *Journal of Biological Chemistry,* **25**, 319–335.

[56] Khabarov, V.N., Boikov, P.Y., Chizhova, N.A. et al. (2009) Comparative analysis of molecular weight distribution of hyaluronic acid in commercial samples (in Russian). *Mesotherapy,* **8**, 18–23. Хабаров, В.Н., Бойков, П.Я., Чижова, Н.А. *и др* (2009) Сравнительный анализ молекулярно-массового распределения коммерческих образцов ГК, *Мезотерапия,* **8**, 18–23.

[57] Khabarov, V.N., Boikov, P.Y., Chizhova, N.A. et al. (2009) The values of molecular weight of hyaluronic acid in the esthetic medicine products (in Russian). *Vestnik of Esthetic Medicine,* **8**, 20–24. Хабаров, В.Н., Бойков, П.Я., Чижова, Н.А. *и др* (2009) Значение параметра молекулярной массы ГК в препаратах для эстетической медицины. *Вестник Эстетической Медицины,* **8** (4), 20–24.

[58] Linker, A, Meyer, K. (1954) Production of unsaturated uronides by bacterial hyaluronidases. *Nature,* **174**, 1192–1194.

[59] Sergiev, V.P., Pal'tsev, M.A. (2008) *Physiology of Parasitism and the Problem of Biological Safety,* Medicina, Moscow, p. 240. Сергиев В.П., Пальцев М.А. *Физиология паразитизма и проблема биологической безопасности.* М.: «Медицина».- 2008 .-С. 240.

4

Molecular and Supramolecular Structure of Hyaluronic Acid

4.1 Primary Structure of Hyaluronic Acid

The molecular structure (chemical structure) of any polymeric substance, that is, its chemical composition and way the atoms are connected in the molecule, does not unambiguously determine the behaviour and properties of biopolymer materials constructed of these macromolecules. The properties of such substances depend also on their supramolecular (physical) structure. This refers to the three-dimensional organization of the macromolecules. The supramolecular structures of the polymeric compounds have various forms that determine the structural and functional properties of biopolymers. It is impossible to observe the structure of biological molecules and their dynamics at the atomic level *in vivo*, though several different physical research methods could be used including hydrodynamic, optical, low-angle X-rays and neutrons diffraction, X-ray structural and neutron-structural analyses, NMR, electron microscopy and scan micro-calorimetry.

It took nearly 20 years (from first discovery in 1934 until publication in *Nature* that included fully described formula) for K. Meyer's laboratory to complete the work that determined the precise chemical structure of hyaluronan macromolecule (Figure 4.1) [1,2]. Molecules of HA are anionic linear heteropolysaccharides built from the regularly alternating residues of D-glucuronic acid and N-acetyl-D-glucosamine. In a molecule of hyaluronic acid, the amino-sugar is attached to D-glucuronic acid by a β-$(1 \rightarrow 4)$-glycoside bond and glucuronic acid with amino-sugar by β-$(1 \rightarrow 3)$-glycosidic bond (Figure 4.1) [3].

All bulky substituents attached to carbon atoms of the pyranose ring (hydroxyl, carboxyl and acetamide groups) and a glycosidic bond with connected hydrocarbon residues occupy an equatorial and sterically more favourable position. The hydrogen atoms are in less sterically favourable axial positions (Figure 4.2), which explains why part of the molecule

Hyaluronic Acid: Preparation, Properties, Application in Biology and Medicine, First Edition.
Mikhail A. Selyanin, Petr Ya. Boykov and Vladimir N. Khabarov.
© 2015 John Wiley & Sons, Ltd. Published 2015 by John Wiley & Sons, Ltd.

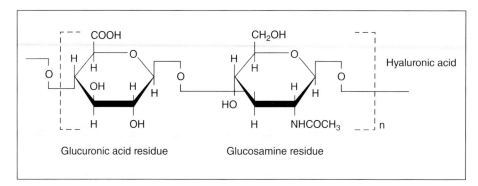

Figure 4.1   *Structure of disaccharide unit of the hyaluronan molecule. Both hydrocarbon residues correspond to β-form of the glucose molecule (β-D-glucopyranose)*

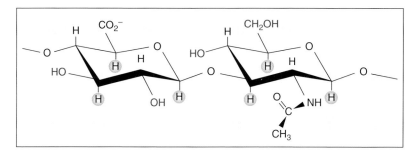

Figure 4.2   *Schematic three-dimensional image of the disaccharide unit of the hyaluronan molecule, which shows the conformation of the hydrocarbon residues*

is hydrophobic. The carboxyl, hydroxyl and acetamide groups provide hydrophilic properties to the polysaccharide molecule.

The presence of polar and non-polar fragments in the macromolecule's structure determines hyaluronan's ability to interact with different chemical compounds and plays a major role in the various conformational transformations. Hyaluronan may exist on the cell surface in an incredibly large number of conformational states (elongated chains, 'relaxed' (non-condensed) helixes, condensed rod-shaped structures, helixes, the structures similar to the pearl necklace and 'clips'). The HA chains, when interacting with each other, form fibrils, webs and other structures.

For the molecule of hyaluronan a characteristic is the creation of a number of hydrogen bonds, which stabilize the macromolecule in aqueous solutions. This has been demonstrated by ^{13}C NMR study [4]. These bonds are formed both inside macromolecule between the neighbouring hydrocarbon residues and the neighbouring molecules. In aqueous solutions the hydrogen bonds could involve water molecules as well. The hydrogen bonds are created between the oxygen of the carboxyl group and hydrogen of acetylamine group either directly or through the molecule of water that serves as a bridge. Other hydrogen bonds are formed between the hydrogen of the hydroxyl group that lie in the equatorial plane and the oxygen atom of the glycosidic bond. It is considered that the created structure

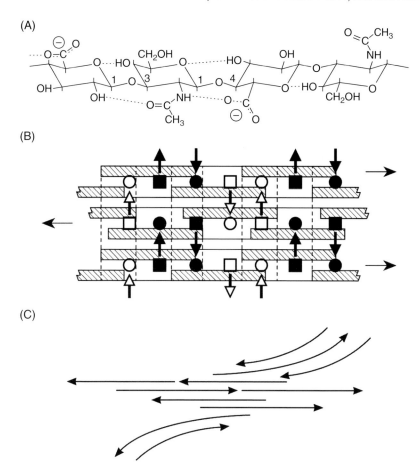

Figure 4.3   *Primary, secondary and tertiary structures in HA solutions [4]. (A)A tetrasaccharide fragment of HA comprising two repeating disaccharides, showing five H bonds that help to maintain the two-fold helix; (B)Three HA chains in a proposed tertiary structure, which accommodates available NMR and x-ray findings, i.e. two-fold helices with gentle curves in the polymer backbone in two planes at right angles (5, 2) and hydrophobic patches (hatched) on alternate sides of the polymer. In antiparallel arrays, the acetamido (■ and □) and carboxylate groups (• and ○) are positioned so that H bonds are possible between them as indicated by arrows pointing from the donor to the acceptor groups; (C) Scheme of overlapping HA molecules that would allow infinite mesh-works to form from HA of high molecular mass. Reproduced with permission from [4]. Copyright © 1999, National Academy of Sciences, U.S.A*

is very stable both in the aqueous solution and in a free state. The Figure 4.3 shows the possible hydrogen bonds and their influence on the HA tertiary structure.

The direct hydrogen bond between the acetamido NH and carboxylate in Figure 4.3(a), seen in dimethyl sulfoxide solution, is largely replaced in aqueous solution by a water bridge. The vertical dotted lines delineate the sugar units and the arrows at the right and left sides indicate the reducing terminal direction. Only in antiparallel orientation do the gentle curves in the backbones of the participating molecules complement each other so that interactions are optimal. As could be seen on Figure 4.3(b), the donor (NH) groups have their NOH bond *trans* to the 2COH bond in this model. Hydrophobic patches (hatched) on neighbouring molecules are contiguous and hydrophobic bonding between them can occur. The H bonds occur in pairs in the proposed structure, with alternating pairs directed in opposite senses (up and down in the diagram). This structure is formally equivalent to that of the beta-sheet in proteins, in which pairs of H bonds are disposed in alternate directions (up and down) between antiparallel polypeptide chains. In combination with the hydrophobic patches, these cooperative interactions would enable large numbers of HA molecules to aggregate specifically, opposed by electrostatic repulsion. This situation is similar to that of the DNA double helix, in which repulsive forces are balanced by hydrophobic bonding and H bonding, although the DNA helix is without the ambidexteran property that permits extensive lateral aggregation.

Theoretical study [5] concluded that conformational behaviour of the hyaluronan molecule and its stability is related of the structure of di-, tri- but mostly tetrasaccharide units as the repeat fragment of HA macromolecule (Figure 4.4).

The length and flexibility of HA macromolecules are significantly variable. It should be noted that the flexibility of the polymer chain is one of the most important characteristics of the material, determining its basic macroscopic properties. Chain flexibility is the ability to change its shape under the influence of the thermal motion of chain branches or under the influence of external fields. This property of the macromolecules is due to internal rotation of the individual parts of the molecule relative to each other, which leads to different conformational states (macromolecular coil, helixes and coils). Stabilization of such biologically active structures strictly follows the laws of thermodynamics in which the properties of the solvent and the presence of various salts are paramount. The nature of the cation has the dominant influence on the conformation of the hyaluronan macromolecule as well as other glycosaminoglycans. The most influent cations are Ca^{2+} and Mg^{2+} [6]. In the solid state HA could show a considerable conformational variability, it could pack itself in a such a way that the neighbouring chains become anti-parallel. The nature of the cation and the level of hydration could influence packing. It was shown [6] that the sodium salt of HA

GlcNAc	GlcA	GlcNAc	GlcA
(N′)	(G′)	(N)	(G)

Figure 4.4 *Schematic representation of the tetrasaccharide β–d-GlcNAc(1–4)β-d-GlcA(1–3)β-d-GlcNAc(1–4)β-d-GlcA, considered as repeat fragment of HA [5]. The GlcNAc residue is abbreviated to as N, whereas the GlcA residue is abbreviated to as G. The labelling of the main atomic positions is indicated, along with the torsion angles of interest*

which does not contain water has elementary cell of the tetragonal form ($a=b=0.989$ nm and $c=3.381$ nm). The hyaluronan chain is a left-hand helix consisting of the repeatable four disaccharide units. In this structure, there are intramolecular hydrogen bonds OH⋯ O through the $1 \rightarrow 3$- glycoside bonds and NH⋯ O through $1 \rightarrow 4$ glycoside bonds as well as the coordination bonds O⋯ Na⋯ O. At a relatively high hydration of sodium hyaluronan we could observe a creation of the elementary cells of orthorhombic forms ($a=1.153$ nm, $b=0.989$ nm and $c=3.380$ nm). The orthorhombic form contains at least four water molecules for each tetrasaccharide unit of the polymer.

Using NMR spectroscopy it was revealed that HA has such three dimensional arrangements of donor and acceptor groups, which is necessary for a creation of stable hydrogen bonds between the acetamide, carboxylate groups [7].

IR-spectroscopy allowed researchers to find that the arrangement of hydrogen bonds in the hyaluronan macromolecule depends on the nature of the cation. Using NMR and HA oxidation with periodate, it was shown that conformational features of the macromolecule related to the hydrogen bonds between the carboxy, acetamide and hydroxyl groups are preserved in aqueous solutions of the polysaccharide [6].

Hyaluronan is an anionic (in the range of the physiological pH values the carboxy group of the D-glucuronic acid residue is practically completely deprotonated), linear polysaccharide, the molecular weight of which essentially depends on the source and the production method (see Chapter 2). Usually the value of the molecular mass is in the range 10^5–10^7 Da, which corresponds to 25 000 and more repeatable disaccharide units.

Other glycosaminoglycans chemically related to hyaluronan, such as chondroitin sulfate, keratan sulfate and heparan sulfate have a much lower molecular weight and have significantly much more variety in the isomers. Such isomers variety is due to presence of sulfated groups, number and arrangement of which could vary significantly. Hyaluronan, isolated from a variety of sources, is always chemically identical, the only difference between the HA products is molecular weight.

4.2   Structure of Hyaluronan in Solution

HA and its salts with the alkali metals, magnesium and ammonium ion are well soluble in water and even in low concentrations form highly viscous aqueous solutions. In high concentrations (1–4 weight %) it forms pseudo-gels: solutions with enormously high viscosity [8]. In neutral aqueous solutions the HA macromolecule is negatively charged. The pK value of the carboxyl group is 3–4 depending on the counter ion - Na^+, K^+, Ca^{2+}, Mg^{2+}, etc.). It is important for the functioning of hyaluronan in the extracellular matrix of the connective tissue.

Hyaluronan salts with cations of two and more valence are practically insoluble in water. Such ions, being introduced into polysaccharide solution, form intermolecular cross-links, which leads to formation of the stable gel structure with high water content. Hyaluronan also forms the salts with inorganic and organic bases that are insoluble in water, for example, with cetylpiridinium chloride as well as meta-chromatic complexes with toluidine blue and with some other organic dyes. Hyaluronan can interact with different proteins. As a result, the complexes with incredibly high viscosity could be formed. The esterification of the carboxyl groups of HA leads to the reducing solubility of the polysaccharide as well [8].

The average longitudinal dimension of the free disaccharide is approximately 1 nm, so the total length of the unfold macromolecules with a molecular weight of 2.5 million Da in solution could be more than 10 μm, which is approximately equal to the diameter of the human erythrocyte and much higher than the average size of a bacterial cell. Obviously, the hyaluronan molecule at sufficiently high concentrations in an aqueous solution takes a more or less compact coiled shape. Formation of coil structure is characteristic for bio-polymers because the necessary condition for formation of such a structure is the level of homogeneity in size of the macromolecules.

By X-ray analysis, laser light scattering and NMR spectroscopy – the methods which were already used for study of hyaluronan aqueous solutions, numerous conformations were found. All of them were related to ion surrounding, concentration, temperature and so on. The macromolecule of HA could band and form left-oriented single and double helixes, or even form multi-strand plane structures [7]. Using X-ray analysis, it was shown that hyaluronan Na and K salts exist as a double helix [9], yet in another publication it was found that the double helix consists of left-oriented anti-parallel strands [10]. Cleland et al. determined the existence of the moving rings formed by hyaluronan molecules with the molecular weight more than 100 kDa at the average ionic strength and the neutral pH of the solution [11]. Using NMR, Darke et al. calculated that part of the rigid structure in the hya-luronan molecule could be up to 50–70% [12]. It could be explained by existence of the rigid segments connected to each other by flexible chain fragments. Using viscometry and laser light scattering methods in combination with X-ray diffraction, the authors [13] found that besides double helix structure, the supra-helical conformation exists as well. In this conformation, HA forms fragments of dense microgel. With increased polysaccharide con-centration in solution, it could be observed a formation of the intermolecular network. In the diluted aqueous solution HA behaved as an independent molecule, but in the high con-centration solutions the web-like structures appear. It happens because of inability of the neighbouring molecules to accept different conformations independently from each other. Only such conformational states are possible, in which every disaccharide fragment of each macromolecule accept the space that is not occupied at that moment by other fragments of the same or neighbouring macromolecule. It means that free HA macromolecules should be rolled into a coil. Such a state is most likely because an extended conformation could be realized by one-way but folded conformation: by many means.

In several publications, it was shown that the secondary structure of HA is similar to plane bands converted into a helix (Figure 4.5) or folded into a sheet [4,7]. The experimen-tal data based on the analysis of changes in NMR chemical shifts and slow oxidation by periodate show that the rigidity characteristic to HA and partial ordering of the HA struc-ture in solution depend mainly on intermolecular hydrogen bonds [2,7,9,10]. NMR study of the oligosaccharides of HA with radioactive labels confirmed that the hydrogen bonds are responsible for atypical secondary structure that the polysaccharide macromolecule can adopt [14].

The additional hydrodynamic studies of the solutions of hyaluronan [9–11] and particu-larly viscosity measurements conducted for diluted solutions show that in the aqueous salt solutions the macromolecule accepts the structure of the statistical semi-rigid coil and this consists of the helix-like bands and the band rings of helical structure [14]. By forming a more or less rigid coil, the molecule binds a large amount of water and creates quite large domains of tertiary structure. Because of the electrostatic repulsion between the negatively

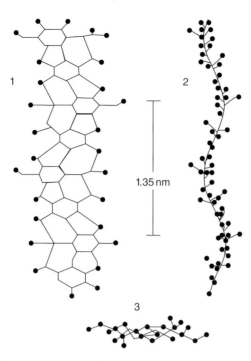

Figure 4.5 *The model of HA structure, based on [4,7]. It is clear that the molecule has the shape of plane band organized into helical structure*

charged groups in hyaluronan, macromolecular super folding does not occur. The real density within the molecular chains of hyaluronan within molecular domain is quite low – about 0.1% by weight. It means that at the concentration of the polysaccharide in an aqueous solution of 1 mg/ml and above, the individual macromolecular coils interact with each other. Thus, only in very dilute solutions macromolecular coils weakly interact with each other, maintaining their individual structures. When the concentration is elevated, the intermolecular structure, similar to the three-dimensional grid, is formed. Such a structure is mainly responsible for the rheological properties of the biological fluids and tissues, such as vitreous [8]. The conformational transitions such as the helix-coil are extremely interesting phenomena characteristic to biopolymer macromolecules such as proteins, nucleic acids and polysaccharides. Such transitions are accompanied by a sharp change in a number of important physical, chemical and biological properties of the system.

The network structure of HA in solution is of great importance. The small molecules, such as water, electrolytes and nutrients can freely diffuse within low-density domain of the biopolymer molecules. At the same time, large molecules, such as proteins, dwell because of their large hydrodynamic size. The decisive factors in the formation of a network structure are the concentration and molecular weight of the polysaccharide. Electron microscopy allows us to reveal that the network structure in solutions of high molecular weight HA (MW of 1 MDa and more) is created, even at very low concentrations. Electron micrographs with rotary shadowing allow visualizing the network structure of high molecular weight HA, which appears endless in solution. The polysaccharides of low molecular weight form a network structure at low concentrations in aqueous solution as well. However, the network has an island-type structure, where some areas are separated by the significant distance [7]. It is important that these networks are highly organized. An electrostatic repulsion of carboxyl groups and the hydrogen bonds between the hydrogen atom of acetamide group and the oxygen atom of carboxyl group in addition to hydrophobic interactions make the antiparallel intermolecular network structure of hyaluronan very favourable. It could be extremely important in the intercellular communication and organization, for example, in the process of morphogenesis. Also, it has been found that a specific structure, which HA molecule could create *in vivo*, depends on hydrodynamic conditions, chemical environment and a presence of other biopolymers, specifically those bound with HA.

The most extended chains, fibres and networks can be formed in the narrow intercellular space, but free ring-like and random coil forms could be found in liquid connective tissues. Consequently, the structure of hyaluronan *in vivo* should be viewed as a product of conformational ensembles and a result of the impact of local environment [8].

Thus, we can conclude that high molecular weight HA in aqueous solution forms a semi-rigid random coil conformation, the time-averaged form of which is the sphere, but every moment the molecule is far from being spherical. This effect occurs as a result of the motion of the segments of the molecule relative to each other at a nanosecond time scale. The dynamic coil of HA molecules is not dense and includes a large amount of water molecules. Even at very low hyaluronan concentrations the intermolecular interaction and ordering takes place. As a result, an intermolecular three-dimensional network structure is formed. Such structure stipulates the unique properties of HA [8].

4.3 Rheological Properties of Hyaluronic Acid

Relatively recently, a couple of decades ago, a new group of oligosaccharide phytohormones were discovered and this substantially broadened and changed understanding of certain aspects of the problem of chemical regulation in plants [15].

Oligosaccharides, being short usually seven or eight units long, branched oligosaccharide chains made of simple monosaccharides (e.g. glucose), appeared to be specific regulators of plant's growth, development, reproduction and activation of different defence mechanisms. In contrast to already known phytohormones that exhibit multiple actions, every oligosaccharide transmits a signal to regulate very particular functions. A similar feature is characteristic to hyaluronan whose multiple biological effects depend on the molecule's molecular weight, that is, molecular size [16–21]. This unique property has not yet been fully scientifically explained but might likely be associated with the self-organization of the macromolecule into a defined liquid-crystal phase. Liquid crystals are the substances that under certain conditions of temperature, pressure and concentration may undergo transition into a liquid crystal state. It is known that biological liquids can be considered as self-organized materials that spontaneously form structures of a liquid-crystal order. There are multiple examples of self-organized biological systems. For instance, the RNA of the tobacco mosaic virus binds to the coat proteins and this results in formation of a rod-like helical virion, capable of infecting tobacco plants. Polypeptide chains of many globular enzymes can fold correctly to form functional biocatalysers. Multiple lipids easily form bilayer structures, which are then spontaneously locked into vesicles.

In view of fact that biological processes are first of all the processes of ordering, their physical interpretation is based principally on the theories of phase transitions and cooperative processes. Biomolecules display properties of lyotropic and/or thermotropic liquid crystallites and exhibit properties that correspond to their liquid-crystal structure. Liquid crystals (nematic liquid-crystal phase) possess unique physical properties. Namely, they combine fluidity of usual liquids with the typical characteristics of crystals. In lyotropic liquids, such properties manifest themselves within a certain temperature range or in a strictly restricted range of concentration of molecules of a given molecular weight (chain length). In this connection, of great interest is investigation of rheological properties of lyotropic biological polymers, in particular, their viscosity (one may say that rheology indeed studies viscoelastic properties of materials). An ability of biological macromolecules to participate in numerous vitally important processes, such as for example, shock damping in cartilage, contraction of cross-striated muscles, skin elasticity and the movement of blood cells in the arteries – in all those processes the investigation of the viscoelastic properties is necessary for understanding biopolymer functioning in actual biological systems. It should be noted that investigation of viscoelastic properties of polyelectrolytes, as hyaluronan belongs to this class of polymers, is a complex problem from both theoretical and practical points of view. Depending on the length, macromolecules form flexible, semi-rigid and rigid chains that affect the orientational ordering of a biopolymer in solution. Under the influence of thermal motion biopolymer chains can adopt different conformations without changes in configuration, that is, without breakdown of chemical bonds. To describe the structure and dynamics of each of these classes in solution, different theoretical models are used. Rigid and semi-rigid polymer chains form liquid-crystal phases. As already noted, the substance molecules in a liquid-crystal state have the property of self-organization, forming spatially organized structures. The organization takes place at different levels of the hierarchy: from nanoscopic to mesoscopic and further to the macroscopic level. All this is in line with the principle of minimum energy conservation in a particular state. There are several types of conformations of macromolecules: macromolecular coil conformation, the elongated rod-like conformation, helical conformation and coil conformation. Helix-type systems seem to be the most energetically favourable. Information about the helical nature of hyaluronic acid macromolecules in an aqueous solution was provided in the previous chapter.

The energy associated with the deformation of liquid crystals is small and, therefore, their molecular structure can easily be changed using external fields, for example, small changes of mechanical pressure or shear flow. Structured fluids are the structures induced by shear. The two main properties of nematic liquid crystals are crystallinity (high-order orientation) and fluidity. Most of the solute–solvent interactions are the interactions leading to a high-order organization. There are several criteria for formation of polymeric nematic lyotropic structures [22]:

- concentration and molecular weight of biopolymer above a critical value;
- temperature below a critical value.

These critical values depend on the type of the solvent. It is known that the nematic liquid crystal phase contains a higher concentration of polymer than does the isotropic phase [23,24]. The liquid crystal state can be associated with chemical structure as well as size of macromolecules.

Rod-like molecules can orient themselves in solution along a direction that leads to appearance of fluid anisotropy and reduction of its viscosity. The changes in pH and ionic strength of a solution can change the size of polyelectrolytes by five or more times. There are different types of supramolecular structure that are capable of reproducing the 'corresponding rigidity' characteristic for liquid crystallites. Some macromolecules, including hyaluronan, have certain conformational flexibility. Conformational motions of flexible parts allows for orientational arrangement of mesogens, which determines the behaviour of a liquid crystal as a whole. Certain flexible and semi-rigid chains such as gamma-benzyl glutamate can become rigid through the spiral-helix transformation. This kind of transformation happens with hyaluronan as well; in this case a phase transition occurs [24]. In diluted aqueous solution macromolecules of hyaluronan behave as independent bodies. In contrast, in highly concentrated solutions the ordered systems resembling those in liquid crystals appear. Crystallization of biopolymer molecules is associated with their flexibility, more precisely, rigidity of macromolecules that cannot be packed otherwise than in certain order, because of limited rigidity and the volume available. Thus, behaviour of the biopolymer macromolecules in solution depends on their rigidity, concentration and, of course, ionic strength and pH of the media. For example, the conformation of hyaluronan is influenced by mutual electrical repulsion between the charges distributed along the polymer chain and this leads to strained conformation of the chains. As the concentration of hyaluronic acid increases, its chains start to overlap, shielding of the charges occurs and the distance between the chain ends decreases so that a network structure is formed. It seems that macromolecules of hyaluronic acid similar to polypeptide molecules can exist in solution in two different forms, that is, ordered state and random coil. Under certain conditions the helix-to-coil transitions, sometimes called 'intramolecular melting', suddenly occur. The peculiarity of such a transition is that this type of intramolecular melting occurs within a single macromolecule and is associated with disappearance of ordered structure. The helix-to-coil transition happens in biological systems in various situations and involves different biopolymer molecules like polysaccharides, polypeptides and nucleic acids. It is important to note that such transitions under natural biological conditions can occur both in one and in the reverse direction, depending on the external stimuli. Biopolymer chains can undergo the spiral-to-helix transition when parameters of the surrounding such as temperature, pH and ionic surrounding are changed; this is of big biological importance. It seems that here one should search for a link between biological properties and molecular weight of hyaluronan.

It is known [24] that liquid-crystalline biopolymers are very sensitive to orientation in external fields, that is, electric and magnetic fields, as well under action of pressure or solution fluidity. When other orienting fields are not present, the dynamics of structured systems strongly depend on the shear stress applied.

The major component of the synovial fluid, that is, hyaluronan, has two main functions; namely lubricant and shock impulse damping [25]. Furthermore, hyaluronic acid is a major component of the extracellular matrix of the connective tissue. This kind of polymer should exhibit viscoelastic properties that directly depend on its microstructure and external parameters such as shear rate, stress and temperature. Knowledge of dependence of model synovial fluid viscosity on the shear rate, stress and temperature is very useful for biomedical applications, for example, in treatment of joint diseases.

Rheology of liquid-crystal polymers is very different from the rheology of usual liquid crystals. However, rheological behaviour of liquid-crystal polymers is different from that of usual polymer liquids too. At the same time, the behaviour of these materials shows some similarities. Namely, liquid-crystalline polymers and liquid crystals behave as non-Newtonian liquids. Their viscosity depends on shear rate. The dynamics of the nematic liquid-crystalline materials in shear flow strongly depends on the molecule positioning and flow direction [26]. Liquid-crystalline polymers can undergo transitions from one state to another. The molecules of liquid crystal relax very rapidly; the macromolecules of liquid-crystalline polymers in contrast relax very slowly, especially in concentrated systems. Intersurface effects (boundary interactions, surface effect, adhesion effects) are very important in describing the close-to-surface dynamics of polymer macromolecules and intersurface viscosity/viscoelasticity [27].

The factors determining secondary structure of macromolecules of hyaluronic acid in the solid state influence its behaviour in aqueous solutions, as it follows from the results of viscometric and other studies. At the same time, macromolecules of hyaluronic acid in aqueous solution may exist in the form of unordered coils that occupy the volume exceeding that of macromolecule chains themselves by more than 100 times. Unfolding of coils is facilitated by weakening of intermolecular interactions under physiological ionic strength and pH [28]. An observation of a sudden yet reversible at pH 7.0 reduction in viscosity at pH 12.5 without changes in the molecular weight demonstrates peculiarities and unique labile nature of hyaluronic acid macromolecules in aqueous solution [5]. Here, we again face the aforementioned spiral-to-helix transition. Upon heating or treatment of hyaluronic acid solution with acid or alkali under mild conditions, sudden changes in hydrodynamic properties are observed. Significant change in solution viscosity occurs without substantial changes in molecular weight of hyaluronan.

The studies by means of NMR and circular dichroism methods [29] made it possible to identify the presence of relatively rigid and more flexible structures alternating within a single hyaluronan chain with a balance between these structures. Extremely high viscosity of hyaluronan solutions was due, on the one hand to its structure and, on the other hand, was conditioned by specific and non-specific intermolecular interactions of macromolecules [29]. It all points to the fact that not only hydrophilic but also hydrophobic interactions play a significant role in stabilization of hyaluronan structure associated with the specific macromolecular structure.

Interactions between hyaluronan macromolecules lead to the formation of intermolecular associations that have a honeycomb structure and occupy an even bigger space than unordered coils. Dynamic properties of such complex structures increase with increasing the ionic strength and decrease with decreasing pH of the solution. Very interesting is another feature: sodium hyaluronan consisting of 3500 disaccharide units in solution quickly forms a spatial structure due to chain interactions, as can be found by viscometry and other methods. Adding to this solution a fraction of hyaluronan consisting of 60 disaccharide units in equal concentrations destroys the structural association. Mixed solutions behave as solutions of isolated non-associated polysaccharide chains. Very short (four disaccharides) and extra-long (from 400 disaccharide units) fragments do not display this particular effect [30]. This is a very important result that must be taken into account when addressing a wide range of issues related to the manufacture of hyaluronan-based medicinal products.

Conformational characteristics of macromolecules define the viscoelastic properties of the various biological macromolecules of biological fluids, extracellular matrix and other structures, formed by this polysaccharide and other proteoglycans. Reduction in the size of polysaccharide macromolecules under pathological conditions can lead to disruption of normal structures, which is dangerous for systems performing optical and mechanical functions (vitreous eyes, synovial fluid, umbilical cord, etc.) and in which hyaluronan is the predominant structural element. By forming a honeycomb structure in solutions the hyaluronan chains bring about the ability of solutions to resist deforming external pressure [31].

By binding significant amount of water and forming a honeycomb structure in aqueous solution, hyaluronan controls the distribution of other similarly charged macromolecules through the steric exclusion from the volume occupied by macromolecules. This effect leads to a concentration of those excluded substances in the extremely limited volume, which may lead to different interactions of these compounds and, consequently, results in the formation of new structures. Hyaluronan as an anionic polymer can interact with low-and high-molecular positively charged compounds. This is most pronounced in the weakly acidic environment, where the interaction of hyaluronan with such substances is seen in the form of insoluble complexes [5]. At the physiological ranges of ionic strength and pH, electrostatic interactions of hyaluronan with proteins and other substances, which do not produce insoluble complexes, may result into a certain space orientation of the molecules and could lead to formation of structures with heterogeneous macromolecular composition.

Matsuoka and Cowman [32] proposed the consideration of two simple models to study the hydrodynamic properties of hyaluronan on the basis of inherent viscosity. The first model of the not freely-jointed or non-ideal coil is based on the statistical conformations of polymer chains. Under this model, the intrinsic viscosity $[\eta]$ is directly proportional to the volume occupied by mass units of the polymer segments. The volume is filled mainly with solvent and has low density of polymer segments. The second model is the freely-jointed chain model. It is assumed that the chain is more extended and its viscosity corresponds to that of the worm-like chains. The intrinsic viscosity changes as a square of the mean-squared end-to-end distance. Experimental studies of intrinsic viscosity of hyaluronan indicate that short chains act as freely-jointed chains and long chain of hyaluronan behave like the not freely-jointed chains. The molecular weight, at which these two types of behaviour co-exist, is approximately 3.75×10^4 Da. The size of the polymer corresponds to a smallest coil of hyaluronan (Figure 4.6).

Thus, it is possible to calculate the molecular mass of hyaluronan based on the viscosity data and vice versa, using the equation that describes macromolecular behaviour of hyaluronan, that is, $[\eta] = 0.029 M^{0.8}$ to the viscosity characteristic for certain molecular weights. This equation is often used for determining the average molecular weight of hyaluronan. On the basis of the results obtained in [32] the conclusion was made that for the macromolecule of hyaluronan in solution being coiled is energetically advantageous. Approximately 20 segments are required to form a coil. The coil (sphere) is of course only a transitory conformational state. The smallest chain length sufficient for a macromolecule of hyaluronan to behave like a coil corresponds to a molecular weight of 37 500 Da.

In the study [33] the effect of urea, sodium chloride, guanidine chloride and sucrose has been investigated. These compounds participate in formation of intermolecular hydrogen bonds and hence affect rheological behaviour of hyaluronan, radius of gyration Rg and

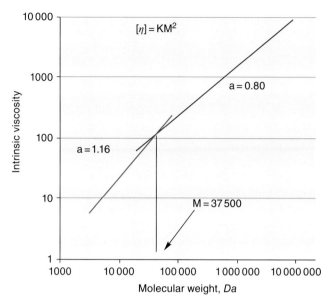

Figure 4.6 *Experimental dependence of intrinsic viscosity on molecular weight of hyaluronan in 0.15M solution of NaCl. Short chains act as freely-jointed chains and long chain of hyaluronan behave like not freely-jointed chains [32]. Reproduced from [32]. With permission from Elsevier*

hydrodynamic radius Rh. Addition of urea does not change either the elasticity modulus or the tangent of dynamic loss, whereas the presence of NaCl and guanidine hydrochloride decreases both parameters and sucrose increases them. The critical concentration C* was calculated as an inflection point on the curves of intrinsic viscosity in hyaluronan solutions containing urea, NaCl, guanidine hydrochloride or sucrose. The changes of all parameters made it possible to assume that sodium and guanidine ions screen electrostatic repulsion between hyaluronan molecules and, consequently, reduce the size of a coil. In contrast, sucrose increases statistical interactions between hyaluronan macromolecules. This agrees well with the light scattering data. The radius of gyration Rg and hydrodynamic radius Rh were found to be the dependent on the concentration of urea and sucrose in the presence of 0.2M NaCl.

Addition of sucrose reduces the size of macromolecule coils and makes hyaluronan molecules more rigid in diluted solutions due to formation of transitory network as a result of hydrogen bonding [34–36]. It is concluded by the authors of those works that even if the breakdown of hydrogen bonds upon addition of urea occurs, this does not impact significantly on the rheological behaviour of hyaluronan. Intermolecular hydrogen bonds that lead to the formation of network in many gel-forming polysaccharides in author's opinion [33] are not realized in hyaluronan solutions. It can be shown, based on the assessment of the linear dimensions of high molecular weight polysaccharide molecules that at a concentration of approximately 1 g/l (0.1% w/w) the macromolecules of hyaluronan almost completely saturate the solution. At high concentrations the molecules overlap, and the solution becomes a network made of hyaluronan chains. The point of the overlap onset is easy to determine; it corresponds to the moment when saturation occurs, that is, the moment after which its viscosity increases sharply with increasing concentration.

Another property of hyaluronan solution, which depends on its concentration, is a shear viscosity [25]. Elastic properties of a hydrogel change as the polymer concentration and molecular weight increase. Fluidity of hyaluronan hydrogel was first determined by C. Jensen and J. Koefoed [37].

By studying the dependence of the viscosity of aqueous hyaluronan solution on the external stimuli, E. Szwajczak [25] provided the isothermal curves for HA_1 and HA_2 samples (Table 4.1), shown in Figure 4.7 and 4.8 at fixed temperatures of 20 and 37 °C, that is, room temperature and the temperature of the human body, respectively.

Figure 4.7 and 4.8 show dependences of η' and η'' on shear rate for polydispersed samples HA_1 and HA_2 at two fixed temperatures.

One may notice that for both materials the viscosity decreases with increasing temperature. In nematic liquid crystal phase, the long axes of the molecule can orient themselves along the fluid flow. It can be assumed that the molecules of structured liquids behave the same way. The viscosity of liquids containing oriented long molecules (especially the long polymer chains) is lower than the viscosity of liquids composed of

Table 4.1   *The samples studied in [17,18]*

Sample	C, mg/ml	Mw	Me
HA_1	10	$5 \cdot 10^5$–$7.3 \cdot 10^5$	$1.8 \cdot 10^6$
HA_1	2.7	$5 \cdot 10^5$–$7.3 \cdot 10^5$	$6.7 \cdot 10^6$
HA_2	8	$6 \cdot 10^6$	$2.26 \cdot 10^6$

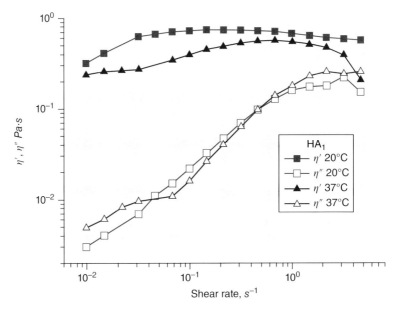

Figure 4.7   *Dependence of viscosities η' and η'' on shear rate for the sample HA_1 at two temperatures: 20 and 37 °C, $M_{w1} = (5 \cdot 10^5$–$7.3 \cdot 10^5)$, $C_1 = 10$ mg/ml*

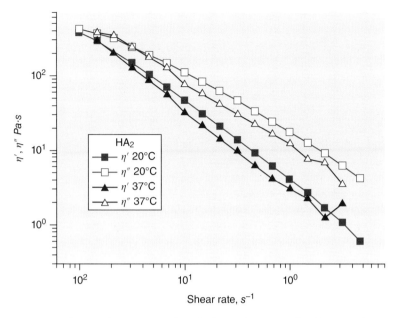

Figure 4.8 *Dependence of viscosities η' and η'' on shear rate for the sample HA_2 at two temperatures: 20 and 37°C, $M_{w2} = 6 \cdot 10^6$, $C_1 = 8\,mg/ml$*

disordered molecules. This means that the viscosity of the material in the liquid crystal phase is lower than the viscosity in isotropic phase [38]. The well-known rheological property of liquid crystals is that the viscosity decreases sharply when moving from isotropic phase to nematic one.

Temperature dependence of the viscosity coefficient η for HA_1 is presented in Figure 4.4. It is seen that the viscosity of the sample initially constantly decreases with increasing temperature. Then, the viscosity increases sharply at a certain temperature. Viscosity decreases again when the temperature is increased, similar to what was already observed. This type of thermal effect becomes noticeable at lower shear rates. This effect is associated with the anisotropic-isotropic phase transition. At higher shear rates this effect diminishes [25].

Another characteristic of liquid crystal polymer systems is the dependence of viscosity on the shear stress τ. Viscosity of complex systems decreases with increasing stress in the range of small shear rates. With further increase in stress the viscosity remains constant in a certain interval of τ and then decreases again.

The values of η' and η'' for HA_2 as a function of shear stress are shown in Figures 4.9 and 4.10. The biopolymer studied shows the behaviour typical for viscoelastic materials with the crystalline order. Viscoelasticity of HA_2 disappears at higher shear stresses. Viscosity η' and η'' significantly decreases at higher shear stresses [25] (Figure 4.10).

By studying the dependence of the viscosity of aqueous hyaluronan solution on its microstructure, Szwajczak [24] has shown the influence of the temperature on the viscosity as a function of shear rate for the two concentrations of the biopolymer (Figure 4.11). The experiments were carried out at different temperatures.

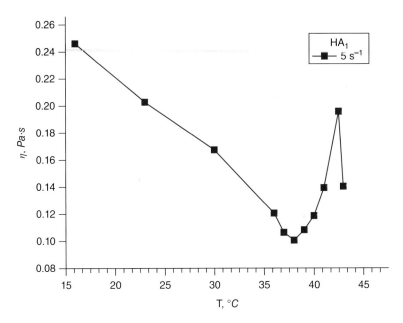

Figure 4.9   *Dynamic viscosity η of the sample HA$_1$ as a function of temperature at 5 s^{-1} shear rate, M$_w$ = (5·10^5–7.3·10^5), C$_1$ = 10 mg/ml*

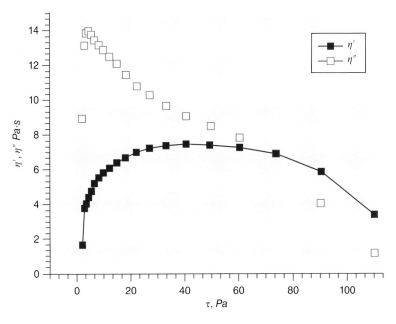

Figure 4.10   *Dependence of η′ and η″ on shear stress for the sample HA$_2$, M$_{w2}$ = 6·10^6, C$_1$ = 8 mg/ml, T, °C = 20°C*

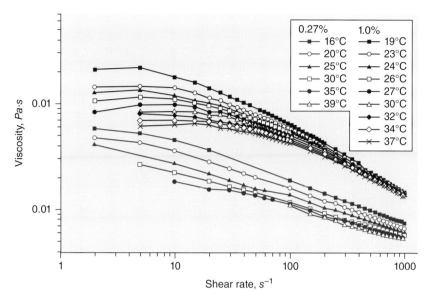

Figure 4.11   *Dependence of viscosity on shear rate at different temperatures for the biological solution of sodium salt of hyaluronan at two biopolymer concentrations (2.7 and 10 mg/ml)*

Figure 4.12 shows the dependence of specific viscosity on the concentration of bacterial HA for molecules having particular molecular weight at fixed temperature at different shear rates. The assessment was carried out on the basis of the data obtained in [39].

$$\eta_{sp} = \frac{(\eta - \eta_s)}{\eta_s}$$

where η_{sp} – specific viscosity, η — solution viscosity, η_s — solvent (water) viscosity.

Viscosity of hyaluronan is the function of shear rate [24,40]. The 'family' of temperature-dependent flow curves for certain concentrations is presented in [24]. It is found that there is a region in the curves where viscosity is almost constant and independent of shear rate. Upon further increase in shear rate the studied solutions display reduced viscosity. Thus, high shear rates favour oriented flows. Sometimes a reduction of viscosity with increasing shear rate may occur when very small rates are applied. Also, a similar effect may occur when the concentrations of hyaluronan is increased [27].

The data on the dependence of viscosity on the concentration, according to Saito et al. [41], show that there is correlation between the concentration c (mg/ml), molecular weight and the point at which entangled coil structure occurs (M_e) in sufficiently diluted aqueous solution, that is,

$$cM_e = 1.8 \times 10^4$$

On the basis of the results listed in Table 4.1, the author [24] comes to a conclusion that the polymers of molecular weight lower than M_e exist as a helix-like structure that can freely adopt various spatial forms. The polymer with the molecular weight higher than M_e arranges itself in a network-like structure that forms a hydrogel due to cross-links. The trend of

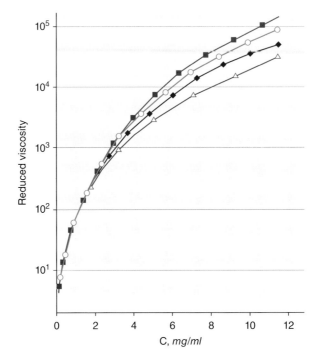

Figure 4.12 Dependence of specific viscosity of bacterial HA on the biopolymer concentration at different shear rates (3; 5; 10; 20 s^{-1}), T, °C = 20 °C, M_w = 2.2 · 10^6 (from Fouissac et al. [39]) (the shear rates increase from bottom to top curves). Reproduced with permission from [39]. Copyright © 1993, American Chemical Society

increasing viscosity can be a direct consequence of chain association. As follows from the previous, viscosity is a parameter that characterizes hydrodynamic properties of macromolecules. It is evident that viscosity also depends on the size and the shape of macromolecules and hence provides insight into the molecule's conformation in studied solution.

The variation of the 'zero-stress viscosity' (also known as specific viscosity η_{sp0} for Newtonian liquids) of HA as a function of the 'rod-likeness' parameter (q) for diluted and semi-diluted solutions at constant temperature is shown in Figure 4.13.

In the study in [42], the influence of different ions of metals on the rheology of hyaluronan solutions at room temperature was investigated (Figure 4.14 and 4.15). It was found that with increasing atomic mass of metal ion there is a progressive reduction of the zero shear viscosity (Williamson viscosity η_0). One can suggest that the drop in viscosity is due to the effect of ions, which are located between the chains of HA and screen electrostatic repulsion between neighbouring carboxylic groups. In addition, the breakdown of the hydrogen bonds between the macromolecules occurs. This results in a more compact coil structure and is accompanied by a reduction in viscosity. This very important property of polyelectrolyte biopolymers is characteristic not only for HA, but also known for polypeptides and nucleic acids and is of paramount importance in biological environments.

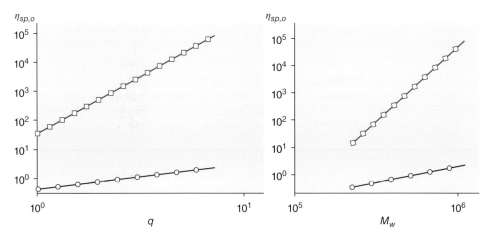

Figure 4.13   *Dependence of viscosity at zero shear rate on the 'rod-likeness' parameter for solutions of bacterial HA in NaCl at the biopolymer concentrations of C=0.5 mg/ml (diluted solution) and C=10 mg/ml (semi-diluted solution); T, °C=25 °C (from the data of Fouissac et al. [39]). Reproduced with permission from [39]. Copyright © 1993, American Chemical Society*

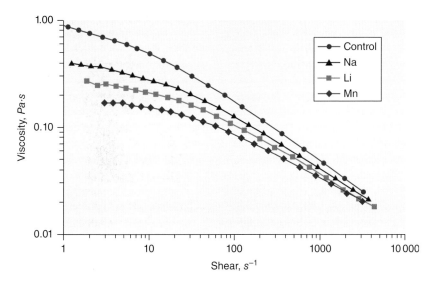

Figure 4.14   *Apparent flow profiles (shear rate versus viscosity) for HA solutions incubated in water (control) and in the presence of Na⁺, Li⁺ and Mn²⁺ ions (all are the $^{Cl-}$ salts) [42]. Reproduced with permission from [42]. Copyright © 2002, Elsevier*

The results presented in Figure 4.14 and 4.15 indeed illustrate intramolecular cross-linking of the biopolymer coils by polyvalent ions. Organism contains large amounts of metal ions (particularly K^+, Na^+, Ca^{2+} and Mg^{2+}), which can change the viscosity and consequently the structure of hyaluronan water salt solutions. The effect of Li^+, the most significant due

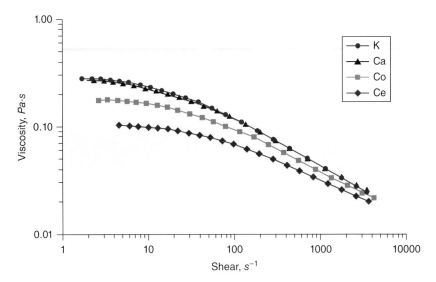

Figure 4.15 *Apparent flow profiles (shear rate versus viscosity) for HA solutions incubated in the presence of K⁺, Ca²⁺, Co²⁺ and Ce³⁺ ions (all are the Cl⁻ salts) [42]. Reproduced with permission from [42]. Copyright © 2002, Elsevier*

to its size and charge, is especially interesting in view of the emergence of a considerable number of pharmaceutical preparations containing lithium.

Cowman et al. [43] investigated the effect of temperature in the range of 25–65 °C on the dynamic rheological behaviour of salt-containing aqueous solution of hyaluronan, Hylan A (cross-linked HA) and a mixture of hylans (known under the trade name Synvisc®). The increase in temperature substantially reduces the modulus and complex viscosity for all three samples.

Thus, in aqueous solutions, HA exhibits the properties of nematic crystalline liquid that manifests itself in its behaviour with the changes in shear rate. Increase in the concentrations of the polysaccharide in solution results in a substantial increase in the volume of macromolecular coils that, because of overlapping, form a hydrogel structure. This structure has the property to be permeable for low molecular weight substances and impermeable for macromolecular compounds. The conformation of HA molecules depends on both the size of the polymer chains (molecular weight) and the pH of the environment as well as the nature of the ions. Conformational flexibility of the macromolecules of hyaluronan determines the appearance of the higher-order structures required for the manifestation of the biological specificity. The possibility of formation of a wide variety of completely or partially ordered supramolecular structures from its macromolecules plays a pivotal biological role in orchestration of intra-and inter-cellular processes.

References

[1] Meyer, K., Palmer, J. (1934) The polysaccharide of the vitreous humor. *Journal of Biological Chemistry*, **107**, 629–634.
[2] Linker, A., Meyer, K. (1954) Production of unsaturated uronides by bacterial hyaluronidases. *Nature*, **174**, 1192–1194.

[3] Radaeva I.F., Kostina G.A., Zmievski A.V. (1997) Hyaluronic acid: biological role, structure, synthesis, isolation, purification, and application (in Russian). *Applied Biochemistry and Microbiology (Prikladnaya biokhimiya i mikrobiologiya)*, **33**, 133–137. Радаева, И.Ф., Костина, Г.А., Змиевский, А.В. (1997) Гиалуроновая кислота: биологическая роль, строение, синтез, выделение, очистка и применение. *Прикладная биохимия и микробиология*, **33** (2), 133–137.

[4] Scott, J.E., Heatley, F. (1999) Hyaluronan forms specific stable tertiary structures in aqueous solution: A 13C NMR study, *Proc. Natl. Acad. Sci. USA*, **96**, 4850–4855.

[5] Haxaire, K., Braccini, I., Milas, M. et al. (2000) Conformational behavior of hyaluronan in relation to its physical properties as probed by molecular modeling. *Glycobiology*, **10**, 587–594.

[6] Bychkov, S.M., Zakharova, M.M. (1979) New data on glycosaminoglycans and proteoglycans (in Russian). *Voprosy medicinskoi khimii*, **3**, 227–237. Бычков, С.М., Захарова, М.М. (1979) Новые данные о гликозаминогликанах и протеогликанах. *Вопросы медицинской химии*, **3**, 227–237.

[7] Scott, J.E., Cummings, C., Brass, A., Chen, Y. (1991) Secondary and tertiary structures of hyaluronan in aqueous solution, investigated by rotary shadowing-electron microscopy and computer simulation. Hyaluronan is a very efficient network-forming polymer. *Biochemical Journal*, **274**, 699–705.

[8] Beloded, A.B. (2008) *Microbiological synthesis and degradation of hyaluronic acid by Streptococcus species bacteria* (in Russian). PhD Thesis. Moscow State University. Белодед, А.В. (2008) *Микробиологический синтез и деградация гиалуроновой кислоты бактериями р. Streptococcus*. Диссертация на соискание ученой степени кандидата биологических наук. Москва

[9] Atkins, E.D., Sheehan J.K. (1973) Hyaluronates: Relation between molecular conformations. *Science*, **179** (4073), 562–564.

[10] Dea, I.C., Moorhouse, M.R. (1973) Hyaluronic acid: a novel, double helical molecule. *Science*, **179** (4073), 560–562.

[11] Cleland, R.L., Wang, J.L. (1970) Ionic polysaccharides. III. Dilute solution properties of hyaluronic acid fractions. *Biopolymers*, **9**, (7), 799–810.

[12] Darke, A., Finer, E.G., Moorhouse, R., Rees, D.A. (1975) Studies of hyaluronate solutions by nuclear magnetic relaxation measurements. Detection of covalently-defmed, stiff segments within the flexible chains. *Journal of Molecular Biolology*, **99** (3), 477– 486.

[13] Ribitsch, G., Schurz, J., Ribitsch, V. (1980) Investigation of the solution structure of hyaluronic acid by light scattering, SAXS and viscosity measurements. *Colloid and Polymer Science*, **258**, 1322–1334.

[14] Almond, A., Brass, A., Sheehan, J.K. (1998) Deducing polymeric structure from aqueous molecular dynamic simulations of oligosaccharides: prediction from simulation of hyaluronan tetrasaccharides compared with hydrodynamic and X-ray fiber diffraction date. *Journal of Molecular Biology*, **284**, 1425–143.

[15] Ovchinnikov, Y.A. (1987) *Bioorganic Chemistry* (in Russian), Prosveschenie, Moscow Овчинников, Ю.А., (1987) *Биоорганическая химия*. Просвещение. Москва.

[16] McKee, C.M., Penno, M.B., Cowman, M. (1996) HA fragments induce chemokine gene expression in alveolar macrophages: the role of HA size and CD-44. *Journal of Clinical Investigation*, **98**, 2403–2413.

[17] Noble, P.W., Lake, F.R., Henson, P.M., Riches, D.W.H. (1993) HA activation of CD-44 induces insulin-like growth factor-1 expression by a tumor necrosis factor-a1-dependent mechanism in murine macrophages. *Journal of Clinical Investigation*, **91**, 2368–2377.

[18] Horton, M.R., Burdick, M.D., Strieter, R.M. et al. (1998) Regulation of HA-induced chemokine gene expression by IL-10 and IFN-y in mouse macrophages. *Journal of Immunology*, **160**, 3023–3030.

[19] Horton, M.R., Shapiro, S., Bao, C., Lowenstein, C.J. (1999) Induction and regulation of macrophage metalloelastase by HA fragments in mouse macrophages. *Journal of Immunology*, **162**, 4171–4176.

[20] West, D.C., Hampson, I.N., Arnold, F., Kumar, S. (1985) Angiogenesis induced by degradation products of HA. *Science*, **228**, 1324–1326.

[21] Ohkawara, Y., Tamura, G., Iwasaki, T., Tanaka, A. (2000) Activation and transforming growth factor-beta production in eosinophils by HA. *American Journal of Respiratory Cell and Molecular Biology*, **23**, 444–451.

[22] Papkov, S.P., Kolontzova, U. (1984) Liquid crystalline order of rigid-chain polymers in *Liquid Crystals Polymers I. Advances in Polymer Science 59*, (ed. N.A. Platé), Springer Verlag, Berlin, pp. 75–102.

[23] Dobb, M.G., McIntyre, J.E. (1984) Properties and applications of liquid-crystalline main chain polymers, *in Liquid Crystals Polymers II/III. Advances in Polymer Science 60/61*, (ed. N.A. Platé), Springer Verlag, Berlin, pp. 61–91.

[24] Szwajczak, E. (2003) Dependence of viscosity of aqueous solution of hyaluronic acid on its microstructure (in Russian). *Rossiiski Journal Biomekhaniki*, **7** (3), 87–98. Швайчак Э. Зависимость вязкости водного раствора гиалуроновой кислоты от ее микроструктуры. *Российский журнал биомеханики*, **7** (3), 87–98.

[25] Szwajczak, E. (2004) Dependence of viscosity of aqueous solution of hyaluronic acid on external fields (in Russian). *Rossiiski Journal Biomekhaniki*, **8** (1), 98–104. Швайчак, Э. (2004) Зависимость вязкости водного раствора гиалуроновой кислоты от внешных полей. Часть II *Российский журнал биомеханики*, **8** (1), 98–104.

[26] Archer, L.A., Larson, R.G. (1995) A molecular theory of flow alignment and tumbling in sheared nematic liquid crystals. *Journal of Chemical Physics*, **103** (8), 3108–3111.

[27] Marrucci, G. (1991) Rheology of nematic polymers, in *Liquid Crystallinity in Polymers: Principles and Fundamental Properties*, (ed. A. Ciferri), Wiley-VCH Verlag GmbH & Co. KGaA, Weinheim, pp. 395–422.

[28] Morris, E.R., Rees, D.A., Young, G. et al. (1997) Order-disorder transition for a bacterial polysaccharide in solution. A role for polysaccharide conformation in recognition between xanthomonas pathogen and its plant host. *Journal of Molecular Biology*, **110** (15), 1–16.

[29] Cowman, M.K., Li, M., Dyal, A., Balazs, E.A. (2000) Tapping mode atomic force microscopy of the hyaluronan derivative, Hylan A. *Carbohydrate Polymers*, **41** (3), 229–235.

[30] Welsh, E.J., Rees, D.A., Morris, E.R., Madden, J.K. (1980) Competitive inhibition evidence for specific intermolecular interactions in hyaluronate solutions. *Journal of Molecular Biology* **138** (5), 375–382.

[31] Scott, J.E., Tigwell, M.J. (1973) Periodate-induced viscosity decreases in aqueous solutions of acetal- and ether-linked polymers. *Carbohydrate Research*, **28** (5), 53–59.

[32] Matsuoka, S., Cowman, M.K. (2002) Equation of state for polymer solution. *Polymer*, **43** (12), 3447–3453.

[33] Noda, S., Funami, T., Nakauma, M. *et al* (2008) Molecular structures of gellan gum imaged with atomic force microscopy in relation to the rheological behavior in aqueous systems. 1. Gellan gum with various acyl contents in the presence and absence of potassium. *Food Hydrocolloids*, **22** (6), 1148–1159.

[34] Szwajczak, E. (2004) Rheological properties of aqueous solutions of biopolymeric hyaluronan. *Proceedings of XV Conference on Liquid Crystals, Chemistry, Physics and Applications*, October 13–17, 2003. Zakopane, Poland.

[35] Nishinari, K., Takahashi, R. (2003) Interaction in polysaccharide solutions and gels. *Current Opinion in Colloid & Interface Science*, **8** (4–5), 396–400.

[36] Hirashima, M., Takahashi, R., Nishinari, K. (2005) Effects of adding acids before and after gelatinization on the viscoelasticity of cornstarch pastes. *Food Hydrocolloids*, **19** (5), 909–914.

[37] Jensen, C.E., Koefoed, J. (1954) Flow elasticity of hyaluronate solutions. *Journal of Colloid Science*, **9** (5), 460–465.

[38] Telega, J.J., Kucaba-Piętal, A., Szwajczak, E. (2002) Liquid crystallinity and temperature dependence of synovial fluid. *Acta Bioengineering and Biomechanics*, **3** (2), 427–434.

[39] Fouissac, E., Milas, M., Rinaudo, M. (1993) Shear-rate, concentration, molecular weight, and temperature viscosity dependence of hyaluronate, a wormlike polyelectrolyte, *Macromolecules*, **26**, 6945–6951.

[40] Fischer, E., Callaghan, P.T., Heatley, F., Scon, J.E. (2002) Shear flow affects secondary and tertiary structures in hyaluronan solution as shown by rheo-NMR. *Journal of Molecular Structure*, **602–603**, 303–311.

[41] Saito, S., Hashimoto, T., Morfin et al. (2002) Structures in semidilute polymer solution induced under steady shear flow as studied by small-angle and neutron scattering *Macromolecules.* **35**, 445–559.

[42] Knill, C.J., Kennedy, J.F., Latif, Y., Ellwood, D.C. (2002) Effect of metal ions on the rheological flow properties of hyaluronan solutions, in *Hyaluronan: Proceedings of an International Meeting, September 2000, North East Wales Institute, UK* (eds J.F. Kennedy, G.O. Phillips, P.A. Williams, V.C. Hascall), Woodhead Publishing Ltd., Cambridge, pp. 175–200.

[43] Cowman, M.K., Li, M., Dyal, A., Balazs, E.A. (2000) Tapping mode atomic force microscopy of the hyaluronan derivative, Hylan A, *Carbohydrate Polymers*, **41** (3), 229–235.

5

Chemical Modifications, Solid Phase, Radio-Chemical and Enzymatic Transformations of Hyaluronic Acid

From the chemical point of view, hyaluronan possesses four different types of functional groups: acetamide, carboxylic acid, hydroxyl and terminal aldehyde. After deacetylation, a free amine could be obtained from an acetamide group. All four functionalities permit characteristic chemical reactions. Such a wide variety of possible chemical modifications creates a sharp difference between hyaluronic acid and other polysaccharides whose reactivity depends mainly upon hydroxyl groups.

We realize that the description of the overall reactivity and chemical modifications of hyaluronan is a dense subject that could fill a separate monograph. However, this chapter will focus mainly on the chemical modifications of hyaluronan that lead to cross-linking. Such modifications play an important role for the creation of hyaluronic acid with valuable chemical and physical properties necessary for biological application of hyaluronan products.

Historically, hyaluronan was known for the chemical transformation of its hydroxyl groups. Only recently, several processes also described transformation of carboxyl groups and deacetylated amino groups. Generally, polysaccharides, including hyaluronan, possess all chemical reactions characteristic to hydroxyl-containing compounds including ethers and ester formation, substitution, elimination, and so on. The reactivity of hyaluronic acids depends mainly on the functionality of hydroxyl groups. Also, the contribution of the aldehyde group is relatively small. The presence of the free hydroxyl groups provides the possibility of structural modifications of the sugar base, which allows direct bio-specific modification to be carried out by means of, for example, bi-functional reagents that interact simultaneously with two functional groups of neighbouring macromolecules. We consider bi-functional reagents as chemical compounds that usually possess two of the same reactive groups separated by a spacer. Bi-functional reagents are widely used for a covalent

Hyaluronic Acid: Preparation, Properties, Application in Biology and Medicine, First Edition.
Mikhail A. Selyanin, Petr Ya. Boykov and Vladimir N. Khabarov.
© 2015 John Wiley & Sons, Ltd. Published 2015 by John Wiley & Sons, Ltd.

linkage of sterically close fragments of polysaccharide macromolecules. One method of bio-specific modification is photo-initiating polysaccharide linkage with an initial introduction of photo-reaction fragments into its structure.

This chapter focuses on two subjects related to the field of mechanically stimulated reactions: (1) chemical and photochemical cross-linking of HA in aqueous solutions and (2) advanced methods of solid-phase polysaccharide modification. From these processes, the cross-linked hyaluronic hydrogels acquire a number of valuable properties that significantly extend the range of their medical applications.

5.1   Main Characteristics of Cross-Linked Hydrogels

In aqueous solution, hyaluronan forms a gel-like structure as a result of intermolecular interaction of linear macromolecules. Colloidal chemistry defines gels as structured systems with liquid dispersion media that exhibit mechanical properties more or less similar to those of solids. The particles of the dispersed phase are connected to each other in a three-dimensional web (which contains dispersion media in its cells and, in the case of hydrogels, water) that deprives the system of fluidity. It is obvious that the properties of hydrogels are mainly dependent on the strength of bonds and level of reticulation in the cross-linked structure. Cross-links in biopolymers can be categorized as either physical (formed by electrostatic interactions or hydrogen bonds) or chemical (formed by covalent bonds). The physical gels, upon heating, undergo web node decomposition that leads to a reduction in the shift modulus. Chemically cross-linked gels are significantly more resistant to heat but at high temperatures a complete and irreversible destruction of the gel's chemical structure takes place. The level of cross-linkage is determined by the average molecular weight of the polymer chain located between cross-links. A cross-link's density directly affects the fundamental properties of the hydrogels, such as degree of swelling, mechanical strength and elasticity, permeability and diffusion characteristics [1].

The hydrogels, formed by hyaluronan as a result of cross-linking, are amphiphilic polymeric substrates able to swell in water and form an insoluble bulky web. The polymeric network is in equilibrium with aqueous environment while there is the balance of elastic forces of cross-linked polymers with osmotic forces of solution. The chemical compositions and the molecular weight of the macromolecule fragment between two cross-links determine a density of the cross-links that, in turn, influences the swelling and size of gel pores [1, 2]. Besides that, the cross-linking characterizes hydrogel as a pseudo-solid compound, not the solution, thereby giving it viscoelastic properties [3]. Such properties are expressed through the physical characteristics of the hydrogels, including swelling level or amount of absorbed water. The swelling is directly related to the chemical structure of the polymer and is in inverse proportion to the density of the cross-links.

In 1943, Flory and Rehner were the first to find a connection between the level of cross-link density of the polymer and its swelling [4]. In the Flory–Rehner model, the swelling level is determined by equilibrium between elastic properties and the forces originated from mixing the polymer and solvent. In 1977, Peppas and Merrill modified the theory of Flory–Rehner to apply it to the behaviour of the hydrogels [5]. Due to elastic forces, the presence of water affects change in chemical potential inside the system [5] and the chemical structure affects the swelling. For example, hydrogels with hydrophilic groups, to which

HA is related, swell more than hydrogels with hydrophobic groups; the latter do not increase in volume in the presence of water [6]. The swelling hydrogel very often may depend on pH, temperature or other factors [7].

The limit of swelling may be determined experimentally or calculated theoretically. The accurate measurement of swelling limit is useful in the calculations of cross-link density, mesh size and the diffusion coefficient. Many natural gels are formed by polyelectrolytes. The physical properties of these systems are greatly influenced by an osmotic pressure caused by counter-ions associated with the polymeric chains. Free counter-ions significantly increase the swelling of the charged gels and affect the elastic modulus. The mechanical characteristics of charged hydrogels determine the properties of many biological structures such as cartilage, synovial fluid, cornea and striated muscle. In order to measure the swelling of hydrogels, different experimental methods could be used. An important feature of hydrogels is the porosity or the size of the mesh. It is a structural property of the material, which is determined as a distance between adjacent cross-links. The study performed with polyethylene glycol diacrylate (PEGDA) experimentally established significant changes in the porosity upon the changes of the polymer molecular weight and small changes of cell sizes at the different concentration [8]. Direct measurements of porosity involve electron microscopy or quasi-elastic laser scattering. Indirect methods include mercury porosimetry and measurements of high elasticity and maximum swelling [7,9].

In designing cellular tissue, the diffusion rate of the solubilized compound is important in order to determine the rate of release of drugs or transport of nutrients and metabolites. The diffusion of nutrients, metabolites and other solubilized compounds depends on many factors, including the morphology of the network, the chemical composition of the hydrogel, the water content, the concentration of solubilized compounds and the level of the material swelling [7].

The nature of the cross-links affects the formation of the hydrogel, its shape, size and degradation. The formation of cross-links should be monitored for the biomedical applications of hydrogels. In this section, three different types of cross-linking in hydrogels are described: covalent, ionic and physical interactions [10].

The appearance of the chemical covalent cross-links can take place during the radical polymerization under exposure to high-energy radiation (gamma and electron) [10]. Before radical polymerization the polymers are usually modified by adding additional reactive groups. For example, acrylate is added to polyethylene glycol (PEG) in order to achieve covalent cross-linking [11]. The radical polymerization of acrylate groups can be initiated by light, high temperature and redox catalysis [12,13].

After the cross-linking process is started, it cannot be cancelled or stopped; it is controlled only by the initial process conditions. Photopolymerization is the conversion of a liquid polymer solution to gel under the action of photosensitizing additives and light [10] and is the most ideal method for synthesis of cross-linked hydrogels intended for use in medical practice since it allows for the reaction to be carried out with almost 100% efficiency.

The second type of chemical bonds in hydrogels represents the bonds based on ionic interactions. Several natural polysaccharides – for example alginate, a natural polysaccharide made from algae, hyaluronan and other charged polymers – form the hydrogel with ionic interaction in the presence of bivalent or multivalent cations. The reaction typically proceeds at ambient temperature and neutral pH [14]. The ionic interactions are weaker than covalent bonds, so such hydrogels undergo rapid degradation in physiological solution (the media in which they are supposed to be used). One example of a synthetic polymer that forms hydrogel with ionic interactions is poly[di(carboxylatephenoxy)phosphazene] [15].

The weakest type of interaction realized by hydrogen bonding, hydrophobic interactions and Van der Waals forces also leads to gel structuring. Hydrogen bonds are usually stronger than hydrophobic and Van der Waals interactions – their energy values are in the range of 10–40 kJ/mol, but they are still an order of magnitude weaker than the ionic and covalent bonds. These weak interactions, however, play a central role in the process of molecular self-assembly, since the various combinations of these interactions in the macromolecules lead to a strong binding. R. Zhang et al. described the molecular self-assembly as a set of molecular building blocks that spontaneously form stable, physically connected network structures [16]. Despite the weakness of each act of physical binding, the multiplicity of such links makes the gel network structures quite stable. Thus, the various chemical and physical bonds can participate in the formation of stable cross-linked hydrogels.

5.2   Methods of Hyaluronic Acid Cross-Linking

As was mentioned previously, hyaluronic acid possesses four functionalities: acetamide, carboxyl, hydroxyl and terminus aldehyde. All are suitable for cross-linkage reactions. Depending on the nature of the cross-linkage reagent, a large variety of hyaluronic acid materials, starting from the films with low water content up to hydrogels with high water content have been synthesized. The majority of the methods of the production of cross-linked hyaluronan is related to one of two schemes: (1) a one-stage process with a bi-functional reagent able to create cross-linked bridges or (2) a two-stage process in which highly reactive HA derivatives are synthesized then followed by second reaction that creates cross-links. Different reagents are typically used for hyaluronan cross-linking including diamines, aminoaldehydes (obtained from aminoacetals), dialdihydes, butadienesulfones, diepoxides, salts of divalent metals and others [17].

5.2.1   Cross-Linking with Carbodiimides

One of the most common reactions of the HA carboxylic group with amino acids and diamines is condensation in presence of HOBT in water/DMSO [18]. It is known that one of the best condensation methods of carboxylic and amino functions is the reaction in the presence of DCC (dicyclohexylcarbodiamide). One of the first studies of the condensation with DCC was conducted in 1991 [19]. A similar method was used for reaction with different amines [20]. Unfortunately, DCC required the reaction in non-aqueous conditions. However, there is a possibility to perform cross-linking with EDC – water soluble analogue of DCC [21]. A similar process, which mentioned condensation with EDC and resulted in cross-linked hydrogels is described in [22,23].

General methods for cross-linking of biopolymers such as hydroxyethylcellulose (HEC), carboxymethylcellulose sodium salt (CMC Na) and hyaluronic acid (HA) using water soluble carbodiamide are summarized in a review published in 2005 [24]. The interesting invention described the synthesis of water-insoluble derivative of hyaluronic acid cross-linked with biscarbodiimide [25].

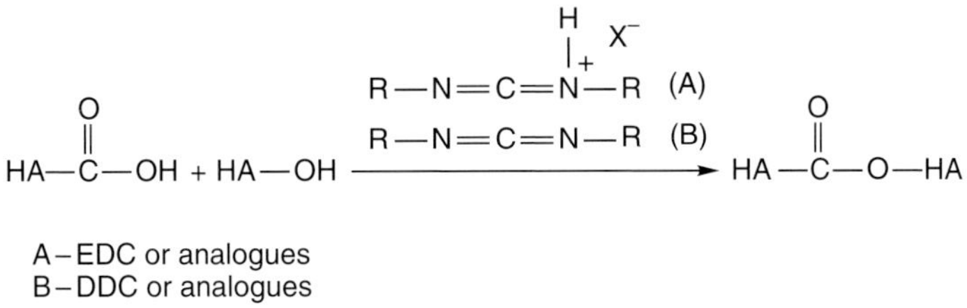

A – EDC or analogues
B – DDC or analogues

Figure 5.1 *Hyaluronic acid cross-linking with carbodiimide*

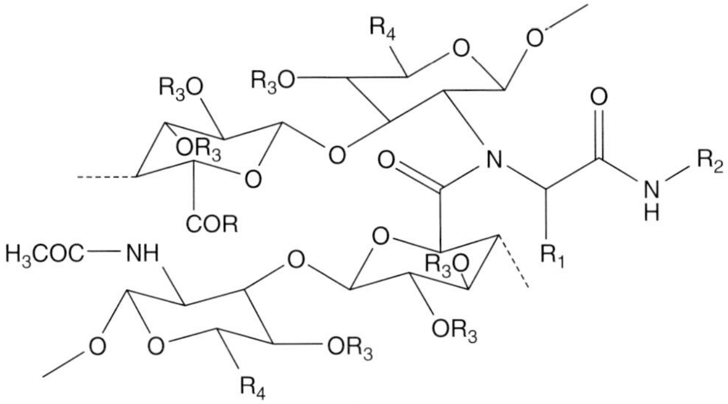

Figure 5.2 *Cross-linked hyaluronic acid with amide bond after reaction with carbodiamide using the Ugi approach*

One of the latest studies investigates the role of the solvent in carbodiimide cross-linking of hyaluronic acid. The cross-linked products were intended for use in ophthalmology. The conclusion was made that after the EDC treatment in the presence of an acetone/water mixture (85:15, v/v) the HA hydrogel membranes have the lowest equilibrium water content, the highest stress at break and the greatest resistance to hyaluronidase digestion. Irrespectively of the solvent composition (in the range of 70–95%), the cross-linked HA hydrogel membranes are compatible with human RPE cell lines without causing toxicity and inflammation [26].

An interesting method of connecting two hyaluronic acid fragments through carboxylic and primary amine functions using the Ugi approach is described in the patent [27]. In the method the primary amine was generated by deacylation, then two molecules were allowed to react in the presence of formaldehyde and cyclohexylisocyanide (Figure 5.2). The Ugi reaction allowed scientists to make cross-linked HA with an N-substituted amide bond, in which carboxylic and amino functions came from the different HA molecules.

Another possible way to synthesize cross-linked HA is by using bi-functional reagents. The cross-linked HA product is synthesized after reaction of HA with dihydrazide in the presence of HOBT and carbodiimide (Figure 5.3) [28]. In this reaction only the carboxylic functions from both HA molecules were used for cross-linking.

Another example of the synthesis of cross-linked HA with a dihydrazide bridge is described in [29]. For a formation of the bond between hyaluronan and primary amine, carbodiimide and N-hydroxysulfosuccinimide were used. The authors synthesized many

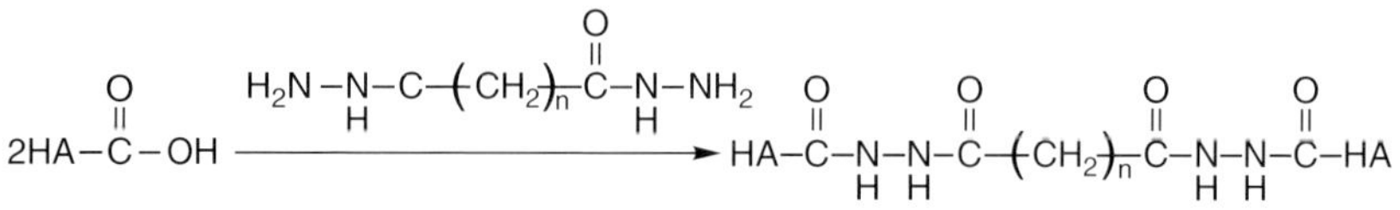

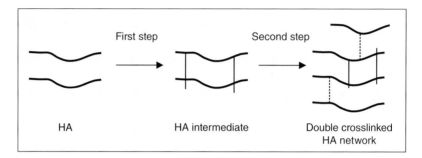

Figure 5.3 *Cross-linking of HA with dihydrazide*

First step   Second step

HA   HA intermediate   Double crosslinked
                                                              HA network

Figure 5.4 *The process of double cross-linking of HA [21]. Reproduced with permission from [21]. Copyright © 1997 John Wiley & Sons, Inc.*

$$2HA-CH_2-OH \xrightarrow{CH_2O} HA-CH_2-O-CH_2-O-CH_2-HA$$

Figure 5.5 *Cross-linking of HA with formaldehyde*

polyvalent hydroxide reagents (2–6 hydrazydes per reagent) for the use in the reactions of HA cross-linking by carbodiimides.

In order to carry out cross-linking, Zhao et al. developed a two-stage method where synthetic polymer polyvinyl alcohol was used at the first stage, followed by ionic biopolymer sodium alginate in combination with hyaluronate (Figure 5.4) in the second stage [30]. This method allowed for the polymer network to be obtained with increased biostability.

5.2.2   Cross-Linking with Aldehydes

Formaldehyde and glutaraldehyde have long been used to cross-link proteins for tissue conservation. Formaldehyde is used for synthesis of the several cross-linked HA products (Figure 5.5). Cross-linking with glutaraldehyde leads to the materials with a high resistance to biodegradation. Tomihata and Ikade studied the reaction with glutaraldehyde in the acidic aqueous solution with acetone with the purpose of obtaining hyaluronic films with restricted swelling [31,32]. Through a comparison of the HA gels cross-linked with glutaraldehyde and carbodiimide, it was found that treatment with carbodiimide leads to final products with a larger amount of cross-link bonds. The obtained materials are used for skeletons of the tissues [33].

5.2.3   Cross-Linking with Divinylsulfone

The stable hydrogels of HA can be obtained by cross-linking of HA with divinylsulfone.

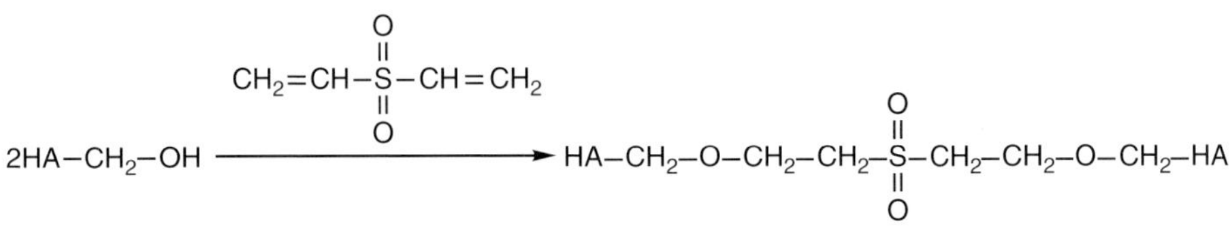

$$CH_2=CH-\underset{\underset{O}{\overset{\overset{O}{\|}}{\|}}}{S}-CH=CH_2$$

$$2HA-CH_2-OH \xrightarrow{\hspace{3cm}} HA-CH_2-O-CH_2-CH_2-\underset{\underset{O}{\overset{\overset{O}{\|}}{\|}}}{S}-CH_2-CH_2-O-CH_2-HA$$

Figure 5.6 *HA cross-linking with divinylsulfone*

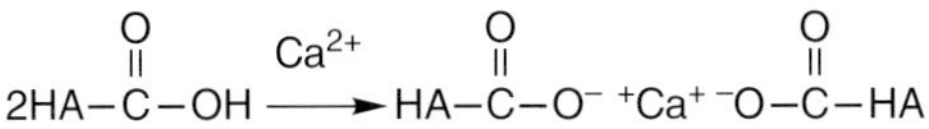

$$2HA-\overset{\overset{O}{\|}}{C}-OH \xrightarrow{Ca^{2+}} HA-\overset{\overset{O}{\|}}{C}-O^-\ ^+Ca^+\ ^-O-\overset{\overset{O}{\|}}{C}-HA$$

Figure 5.7 *HA cross-linking with bivalent metal ions*

In alkaline media divinylsulfone or butadienesulfone, sulfonyl ether linkage is formed between hydroxyl groups of HA. Depending on the reaction conditions, the reaction products could have different consistencies from the soft gels up to solid films (the same method could be used for formation the membranes and hollow pipe). Such products, used as microimplants, are allowed to stay in the body for a long time, especially in the places where implants are not affected by mechanical forces.

5.2.4 Cross-Linking by the Ions of Polyvalent Metals

Adding polyvalent metal salts into the solution of hyaluronan leads to physical cross-linking of polysaccharides due to the formation of ionic bonds. For example, divalent metals such as calcium, zinc, copper and so on, form cross-linked HA salts [34]. However, such ionic binding with polyvalent metals (Figure 5.7) is significantly weaker compared to strong chemical covalent bonds. Stability of such ionic binding depends on different factors such as pH, media, ionic strength and temperature. So, when such hyaluronic acid salts are used as biomedical materials, for intradermal injection in mesotherapy for example, their presence in derma has a short life span and it is difficult to control in order to assure the necessary physiological action of HA hydrogels on the organism [35].

5.2.5 Cross-Linking with Epoxides

Almost 40 years ago, Laurent et al. obtained cross-linked hyaluronan hydrogels using diglycidyl ether of polyetheleneglycol [36]. Other researchers extended the possibilities of diepoxyde chemistry using ethelene glycol diglycidyl ether as a bi-functional reagent and glycidyl ether of polyglycerol as a trifunctional reagent. The cross-linking method by epoxides is described in [37]. The detailed study of the reaction mechanism showed that at the high pH diepoxides form ester bonds involved carboxylic groups, while at low pH ether bonds are formed between hydroxyl groups.

The cross-linking method with diepoxide and epoxide oligomers has great potential because the products, especially in combination with ether links, possess relatively high resistance to hydrolytic decomposition. It is related to the reaction products of diepoxide of bis-alcohols and ethylene glycol oligomers including bis-ethylene glycol [38]. One of the most common methods for HA cross-linking is using diepoxide of bis-alcohols, particularly

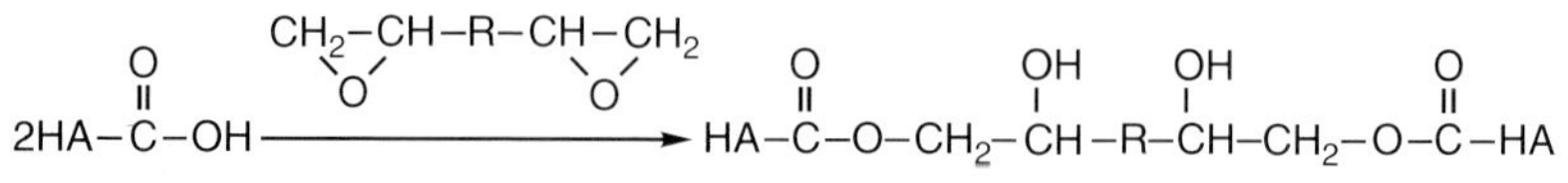

$$\text{2HA-C-OH} \xrightarrow{\quad \text{CH}_2\text{-CH-R-CH-CH}_2 \quad} \text{HA-C-O-CH}_2\text{-CH-R-CH-CH}_2\text{-O-C-HA}$$

Figure 5.8 *Cross-linking of HA with BDDE at the carboxylic group (esterification)*

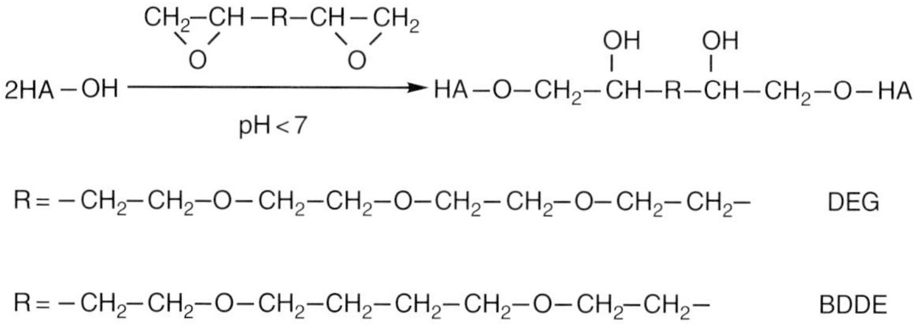

$$\text{2HA-OH} \xrightarrow[\text{pH}<7]{\quad \text{CH}_2\text{-CH-R-CH-CH}_2 \quad} \text{HA-O-CH}_2\text{-CH-R-CH-CH}_2\text{-O-HA}$$

$$R = -CH_2-CH_2-O-CH_2-CH_2-O-CH_2-CH_2-O-CH_2-CH_2- \qquad DEG$$

$$R = -CH_2-CH_2-O-CH_2-CH_2-CH_2-CH_2-O-CH_2-CH_2- \qquad BDDE$$

Figure 5.9 *Cross-linking of HA with BDDE at the hydroxylic group (etherification)*

1,4-butandioldiglycidyl ether (BDDE) in alkaline media [39]. However, this method has several deficiencies:

- Usage of reagent excess, which makes it difficult to predict the level of cross-linking and purification of the final product;
- The reaction takes place using the carboxylic groups, which leads to products which are less stable for hydrolysis and easy degraded in the body (Figure 5.8).

Another HA cross-linking method with BDDE was described by De Belder and Malson [40]. The reaction of HA with BDDE could be carried out in acidic media, which was created by hydrochloric acid. However, it was proved that in acidic media the cross-linking reaction takes place using the carboxylic group as well. The comprehensive study of the BDDE cross-linking in different conditions was performed by Schante et al. [41]. It was finally proved that contrary to early data, the reaction in acidic conditions leads to the cross-linking at carboxylic groups (esterification). The cross-linking in alkaline conditions (pH around 10) takes place at hydroxylic groups (esterification), which leads to more stable products (Figure 5.9).

It is important to mention that cross-linking agents such as carbodiimes, aldehydes and epoxides, are toxic compounds. Their presence in the final medicinal product, even in small amounts is absolutely unacceptable. Purification of the cross-linked hyaluronate gel from technological impurities is a very difficult process that leads to considerable increase in the processing cost of goods. That is why it is important to develop technologically 'clean' cross-linkage processes based on the physical methods of the chemical reactions, such as solid-state modification and photo-cross-linking.

5.2.6 Photo-Cross-Linking

A formation of cross-linked hyaluronan could be achieved by UV-exposure. This is how two covalently linked hydrogels were produced and used in tissue engineering [42]. A modification of hyaluronan and other similar polymers by UV exposure with added photo-active molecules increases the biochemical functionality and mechanical stability of the hydrogels. The molecules of polyethelene glycol acrylate (PEGA) undertook radical pho-topolymerization in order to form cross-linked hydrogels [43].

In order to initiate photo-radical reaction of the chemical cross-linking in HA, its functional groups undergo structural modifications. Leach et al. carried out the reaction between the methacrylic esters and hyaluronate [44]. Shu and Prestwich [45] successfully synthesized HA thioates. Raeber et al. were the first to modify PEGA with the fibronectin derived integrin-binding peptide Ac-GCGY*GRGDSP*G in order to increase cell adhesion and tissue growth stimulation [46]. Similar transformations were carried out with hyaluro-nan for subsequent bone engineering applications, namely to enhance cartilage repair [45].

It is important to point out that it is possible to carry out photochemical transformations in mild conditions in order to maintain the activity of the biological molecules or encapsu-lated cells. For such purposes the photo-transformation of HA is carried out in the presence of several compounds, such as cinnamic acid derivatives, coumarin, thymine, methacrylic anhydride, glycidyl methacrylate and styrene. Figure 5.10 shows how methacrylic esters undergo photopolymerization in the presence of HA to form a grafted polymer [46].

The advantage of such method is that the product of HA photo reactive derivatization is a water-soluble compound. Before the process of photo-solidifying starts, when the three-dimensional web has not yet formed, unreactive toxic compounds with low molecular weight could be easily removed from the reaction zone.

Photo solidifying is thus the last step of a two-stage method of gel cross-linking that takes place instead of the reaction with chemical reagents. At the same time, the first stage includes two operations – the introduction of spacer followed by a photoactive group [35]. As a result, UV exposure permits cross-linked hydrogels to be obtained, which are impurity-free and ready to use. At the same time, this method, as all processes are carried out in the liquid phase, does not ensure equal density of cross-linkage in the whole material volume. This is related to the fact that all synthetic polymers are dispersed by the molecular weight and, in other words, represent a mixture of macromolecules of different sizes.

The natural biopolymers, to which HA belongs, are not polydispersed polymers due to the matrix nature of their synthesis. The nature of the biochemical synthesis is determined by the matrix; the enzyme upon which the biopolymer is synthesized. Nevertheless, during biopoly-mer extraction and purification processes they degrade in one way or another. For example, polygalactomannan, different types of cellulose (wood or cotton), chitosan and hyaluronan are isolated as a wide range of the relatively narrow dispersed macromolecule fractions.

The purification and isolation of the final product of the biochemical synthesis, as a rule, always provides a multimodal set of narrow fractions (multimodal type distribution is a mode in which more than one peak appears on the graph of the molecular weight distribution). This leads to the fact that during the binding of a polydisperse system under mild conditions

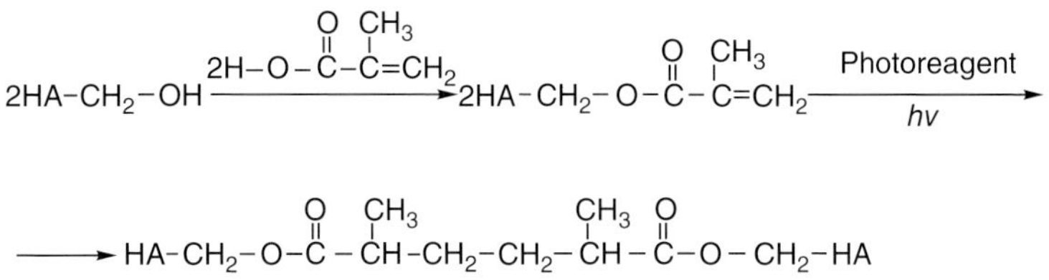

Figure 5.10 *Cross-linking of HA by photopolymerization*

in the aqueous media, the high molecular mass fractions are cross-linked faster than low molecular mass fractions. The result is the obvious problem of heterogeneity of the gel, which is especially noticeable at high degrees of cross-linking. These disadvantages can be avoided by using solid-phase modification process polysaccharides (solid-state reactive blending (SSRB) [47].

5.2.7   Solid-State Cross-Linking under High Pressure and Shear Deformation (Solid-State Reactive Blending: SSRB)

Since the 1970s, systematic studies of the effects of high pressure and shear stress on solid organic compounds have been conducted. The main instrument used in the studies was the Bridgman anvils, the device invented by Nobel Prize winner P.W. Bridgman [48] for studying the detonation properties organic explosives. During that period a number of observations on the evolution of certain organic substances under high (1.5 GPa) pressure and shear deformation were published. More than 40 years ago, Enikolopov published the first report regarding the polymerization of several solid-state vinyl monomers under the effect of high pressure and shear stress. Since then the conversions of hundreds of organic compounds and different types of chemical reactions taking place in these conditions have been studied [49]. Many chemical reactions involving both low molecular mass substances and polymers had been carried out by way of direct and reversed etherification and amidation. These reactions involve epoxides, various types of rearrangements, polymerization and polycondensation and modification of polymers of different structures. The rules identified in these studies demonstrate a critical role of deformation in determination of the conversion degree and existence of critical pressure, below which the reaction does not occur at all degrees of deformation.

The identification of the active states during the different conditions of the mechanical action permits establishment of possibilities to carry out chemical processes and make a prognosis for how to use them effectively. Deformation of solid materials, regardless of their chemical nature, is accompanied by deep disordering of the solid compound through the creation of nano-sized structures and a large amount of active centres (electron- and vibration-exited bonds, electrons and ions stabilized in the traps, low-coordinated atoms in the dislocation nucleus and other structural defects, meta-stable atoms etc.). This is why the solid state mechanical processes are quite similar to photochemical and radiochemical processes.

The basis for solid phase reactions is an intensive mechanical treatment of the solid mixture of the reagents that takes place in the absence of the solvents under the mutual action of high pressure and sheer deformation in different apparatuses. As a result of the treatment, grounding and amorphization of the polymers takes place as the components approach each other at the nano level with the mutual plastic deformation under pressure. It provides non-dislocational mass transfer, which resulted in a deformational mixture at the molecular level [31]. At the end, this leads to high conversion rates in mechano-stimulating chemical reactions. The activation of solid compounds at the high pressure and shear tension creates possibility for the targeted chemical transformations on the non-melting and insoluble compounds with high yield. While the plastic is fluid under pressure, both crystalline and amorphous polymers are affected with structural changes. At the same time the structure is formed, which contains the liable nano-sized areas of the ordered crystalline domains and newly-formed nano-areas of amorphous phase. These structures are highly ordered due to the chain orientation in the flow plane. In cases wherein the cross-linking occurs at local flow (at the level of the nano-sized plastic deformation), all bond systems that could be within this volume are involved into new bond formation. The reason is that the linear size of any macromolecule is significantly bigger than the deformation size (3–5 nm) [47].

Thus, the macromolecules are in the extended conformation and are linked locally during the cross-linking caused by the pressure and shear deformation in the solid-state. Consequently, the HA chains cross-linked in such a mode are more uniform throughout the whole macromolecule length. Besides that, liquid phase cross-linking provides more intermolecular bonds and less intramolecular ones because the macromolecules exist as coils in the solution. In other words, the system cross-linked in the liquid media is less homogeneous in comparison to a system cross-linked in the solid state. By preserving the reaction selectivity, the optimum combination of the temperature and deformation intensity is allowed to reach the high level of the transformation of the functional groups.

The linking agent, which is used in plastic deformation, is distributed through the whole polymer volume at a level close to monomolecular one. As a result, it is possible to obtain cross-linking of any frequency. The method of gel-sol analysis based on the statistical theory of the formation of the web-type polymers allows for the parameters of such network to be determined, that is, concentration and functionality of cross-links and chain dimensions between cross-linking. Such a web structure has nano periods both along the cross-linked macromolecule (5–30 nm) and across the cross-link bridges, which are limited by the length of the linking molecule (usually 2–4 nm).

The method of hyaluronic acid solid-state cross-linking (SSRB) includes initial homogenation of powder HA salt with the linking agents (e.g. diglycidyl ethers of the different diols) [50–53]. The homogenation usually takes place in the mixer at 20–50°C. The mixture obtained is then treated with pressure and sheer deformation on the Bridgman anvil. To do this, the reaction mixture is placed between two cone-shaped anvils. The anvils are pressed to such extent that the sample is subjected to a pressure of up to 10 HPa. After the pressure has applied, the sheer deformation of the sample takes place due to rotation of lower anvil at the specified angle. The treatment of both pressure and sheer deformation has a duration of 2–3 min. The resulting cross-linked HA was also studied by IR-Fourier spectroscopy, NMR methods and X-ray analysis [47,52,53].

The influence of the solid-state cross-linking method on the molecular structure of hyaluronan was studied by the X-ray diffraction method [50]. The study established that all diffractograms of the cross-linked samples of the polysaccharide with two links on every 100 HA units are identical to the diffractograms of the samples of the starting material. The conclusion is that the native molecular structure of HA is maintained at a low cross-linking level. At the higher cross-linking level (8 links on every 100 disaccharide units), the molecular structure of the starting HA has changed slightly: at the 20° scattering angle, the half-width of the more intensive reflex (which is indicative of the crystalline structure of HA) is increased by up to 10%. This data shows an insignificant defect of the molecular structure. The Fourier-transform IR spectroscopy study of the initial HA samples and the cross-linked HA with different cross-linking levels (2–16 links on every 100 units; see Figure 5.11) shows complete absence of the characteristic absorption bonds of an epoxy group (750–950 cm^{-1} and 1240–1260 cm^{-1}), indicating full conversion of the epoxy compounds. After cross-linking completion it is possible to calculate a number of cross-links on each 100 macromolecule units based on the ratio of the reagents in the reaction mixture taken for

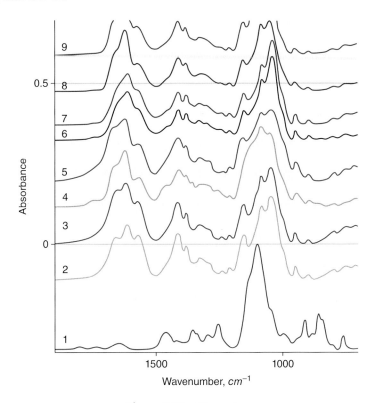

Figure 5.11   *IR spectra:*

1. *Diethelreneglycol diglycidile ether*
2. *Hyaluronic acid sodium salt, anhydrous*
3. *Hyaluronic acid sodium salt*
4. *Hyaluronic acid cross-linked with DEG-1 (100% mol)*
5. *Hyaluronic acid cross-linked on the Bridgeman anvil*
6. *Hyaluronic acid cross-linked with DEG-1 (16 mol%, reflectance mode)*
7. *Hyaluronic acid cross-linked with DEG-1 (8 mol%, reflectance mode)*
8. *Hyaluronic acid cross-linked with DEG-1 (8 mol%, transmission mode)*
9. *Hyaluronic acid cross-linked with DEG-1 (16 mol%, transmission mode)*

cross-linking. The intensity of the characteristic bond of the acetamide bond ($1640\,cm^{-1}$) is equal in the spectra of all samples, but at the same time the characteristic bands of the amino groups are not found. The characteristic ester bands ($1730–1750\,cm^{-1}$) are also not found, which indicates the absence of cross-links through carboxyl groups. It is thus safe to conclude that cross-linking takes place by interaction of epoxy groups of the linking reagent with free hydroxyl groups of hyaluronan. IR spectroscopy cannot answer which groups are involved in cross-linking, which is why the more informative method C^{13} NMR spectroscopy has been used [47] (Figure 5.11, 5.12). This method showed that the decreasing intensity of the C^6 peak (chemical shift 63.5 ppm) is responsible for the carbon atom connected with the primary hydroxyl group.

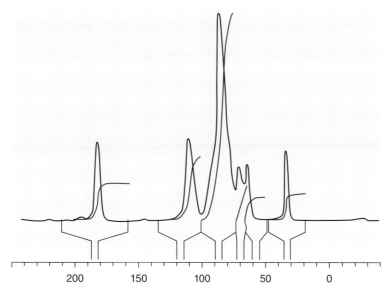

Figure 5.12 ^{13}C-NMR spectrum of hyaluronic acid reference material

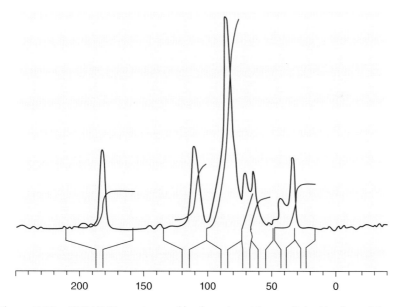

Figure 5.13   13C-NMR spectrum of hyaluronic acid cross-linked in the solid state

Experimental data shows that the process of solid state cross-linking performed in a single technological regime through the simultaneous action of pressure and sheer deformation creates equally cross-linked samples of hyaluronan with pre-designed levels of cross-linking without foreign technological additives. Several advanced results have been obtained through methods of mechano-chemical action. Many chemical reactions that involve both low- and high molecular weight polymers have been studied, including direct and reversed esterification reactions, amidation, different types of rearrangement and different types of polysaccharide modification. The mechanical activation of the substrate and reagent through their intensive mixing allow for the performance of processes not possible in liquid media. The developed methods of solid-state synthesis and modification of polysaccharide look extremely prospective for obtaining the materials of biomedical application. The main advantages are exclusion of the toxic solvents, reaction initiators and catalysts. Many unique products could be obtained using solid-state modification of the biopolymers (see Chapter 6). Such products are difficult or may be even impossible to synthesize by any other method known in the chemistry of polymers.

5.3 Radiochemical Transformations (Radiolysis) of Hyaluronan Aqueous Solutions

A considerable number of published studies on hyaluronan are related to studies of the chemical transformations of the polysaccharide by the action of ionized radiation. Radio-chemical transformations of hyaluronan aqueous solutions are studied by the radiation chemistry of biopolymers. The interest in this research is related to the practical problems of radiobiology: the protection of living cells and biological media against ionizing radiation and, on the contrary, stimulation of their decomposition (e.g. in the case of cancer radiation therapy). Secondly, methods used in radiation chemistry such as pulse radiolysis, electron paramagnetic resonance (EPR spectroscopy) and others, can provide valuable quantitative information about the reactivity of different radical particles related to the macromolecule of hyaluronan. Thirdly, the effects of ionizing radiation can be considered as a method of physically stimulating polysaccharide chemical reactions for subsequent modifications in order to provide unique properties of the material. Finally, knowledge of the processes occurring during irradiation optimizes radiation sterilization of hyaluronan medical products.

During the initial stages of irradiation of hyaluronan aqueous solutions, the main effect is the reduction of the viscosity related to the depolymerization process [54–59] in which glycosidic bonds are hydrolysed and low molecular weight oligosaccharides are formed. With the increase of the absorbed doses of electronic and gamma radiation, a deep degradation of HA macromolecules and the destruction of pyranose ring takes place [60–64]. Table 5.1 shows data related to reduction of the average molecular weight depending on the absorbed dose for four samples with different initial HA molecular weights.

The data presented in Table 5.1 show that increase of the absorbed dosage of gamma-irradiation exposure resulted in a sharp decrease of the molecular weight of all HA samples. At the same time the fraction composition of the exposed polysaccharide did not change. Similar results were obtained after irradiation of another similar polysaccharide: carboxymethylcellulose [61,63]. These data show that the radical particles that originated at the exposure of the aqueous solutions effectively react with the macromolecule of the polysaccharide, which leads to its degradation. The action of the ionized exposure on water resulted in formation of extremely active radical particles, such as hydroxyl radical (OH*), atomic hydrogen (H*) and hydrated electron [65]. The hydroxyl radical removes the hydrogen atom from the polysaccharide macromolecule, while hydrated electrons mainly attack carbonyl groups, which results in the formation of anion-radicals. The radical OH* could

Table 5.1   *Dependence of the average molecular mass (MW) and molecular mass distribution (MW/MN) on absorbed gamma-irradiation dose [61]*

Dose, kGrey	Lot A MW, kDa	Lot A MW/MN	Lot B MW, kDa	Lot B MW/MN	Lot C MW, kDa	Lot C MW/MN	Lot D MW, kDa	Lot D MW/MN
0	1640	1.1	1800	1.2	1100	1.3	2100	1.1
2.0	420	1.2	430	1.2	340	1.2		
4.1	260	1.2	290	1.2	210	1.2		
6.4	180	1.2	190	1.2	150	1.2		
7.8	170	1.2	180	1.1	130	1.1		
8.9							260	1.2
11.0	130	1.2	150	1.1	130	1.1		
11.9							180	1.1
12.9	150	1.2	130	1.1	110	1.1		
14.3							150	1.1
14.9	110	1.1	110	1.1	92	1.1		
17.1	100	1.1	100	1.1	81	1.1		
18.1	80	1.2	90	1.1	75	1.1		
20.4							120	1.1
26.2							90	1.1
30.0	78							
60.0	44							
90.0	29							

attack all carbon atoms in the carbohydrate cycle, to which hydroxyl groups are attached; or, carbon atoms involved into glycosidic bonds. In the first case, unpaired electrons localized on any 'middle' carbon atom of the pyranose cycle, in the second case they localized on the atom C_1 that forms at the cleavage of the glycoside bond in the position 1–3 or 1–4. As a result of such attack the polysaccharide macroradicals of different structures are formed [66]. Using pulse radiolysis, the rate constants for the reaction of HA with hydrated electron e^-_{aq} and OH* radicals: $K(e^-_{aq} + HA) = 3.5 \times 10^8$ $M^{-1}s^{-1}$ and $K(OH* + HA) = 8.4 \times 10^8$ $M^{-1}s^{-1}$ were determined [67]. From the analysis of data presented in [68–71] for radiation-chemical decolouring of different compounds (including HA), it is possible to make an estimation of the value of $K(OH* + HA)$ in a competing reaction of OH* radicals with dye. Such estimation leads to the value of $K(OH* + HA) = (1.0 +/- 0.5) \times 10^9$ $M^{-1}s^{-1}$, which correlates well with the reported data [65]. Similar rate constants of the OH* reaction and hydrated electron with heparin, keratan sulfate and carboxymethylcellulose were presented in [72].

It is worth mentioning that if the aqueous solutions saturated with air are exposed by ionized irradiation, additional particles O_2^- and CO_2^- are formed that participate in polysaccharide radiolysis as well. The majority of researchers confirm that the reactivity of the radical particles formed during irradiation decreases in the series $OH* > e^-_{aq} > CO_2^- > O_2^-$ [73].

Thus, based on the reactivity of the different radicals, it is possible to conclude that the destruction (reduction of the molecular weight) of HA during aqueous solution takes place mainly as a result of the reaction of polysaccharide with the oxidative radical OH*. The reactions of the reductive radical agents e^-_{aq}, CO_2^- and O_2^- are less efficient compared to OH*, however, they could transform the acetamide group into an amino group without effecting the main polysacchride chain (Figure 5.14).

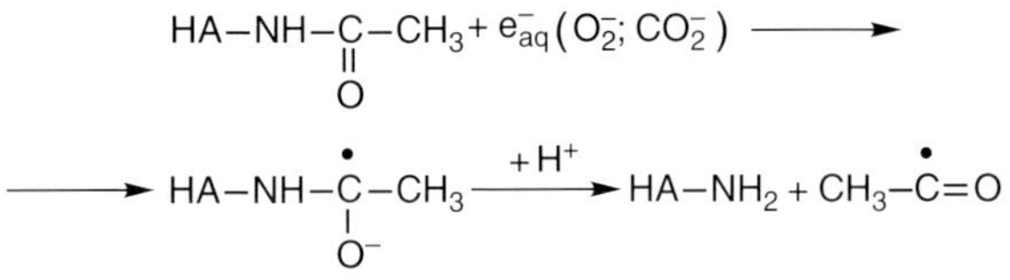

$$HA-NH-\overset{\displaystyle O}{\underset{\displaystyle \|}{C}}-CH_3 + e^-_{aq}(O^-_2; CO^-_2) \longrightarrow$$

$$\longrightarrow HA-NH-\overset{\displaystyle \bullet}{\underset{\displaystyle |}{C}}-CH_3 \xrightarrow{+H^+} HA-NH_2 + CH_3-\overset{\displaystyle \bullet}{C}=O$$

Figure 5.14 *The reaction of the electron transfer during interaction of the reducing radical with the HA macromolecule*

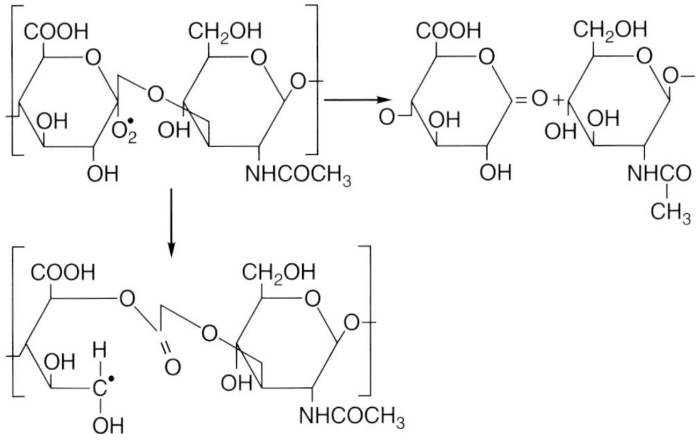

Figure 5.15 *Different paths for the transformation of the peroxide hyaluronan macroradical [62]. Reproduced from [62]. With permission from Elsevier*

The exposure of hyaluronan solutions in presence of solubilized oxygen leads to a rate increase of several destructive radiation and chemical transformations of the polysaccharide [57,62]. The molecular oxygen influences the yield of the primary radical and final molecular products. The formation of such products is caused by the deep destruction of the pyranose cycle. The macroradicals of hyaluronan formed upon irradiation react with the oxygen molecule with formation of two peroxide type radicals. The structure of these peroxide macroradicals was established by EPR-spectroscopy [55,64,66]; the first is formed as a result of the reaction of O_2 with a free fragment at the atom C_1, the second, less stable, is formed upon attack of the molecular oxygen of the non-paired electron at the 'middle' carbon atom of the pyranose cycle. Then part of the peroxide radicals ROO* is decomposed following the reaction, as shown in Figure 5.15 [74], which leads to destruction of the polysaccharide macromolecule. Another part is participated in the reaction with undamaged HA molecule, which results in formation of hydroperoxide compounds of the general formula ROOH. The following decomposition of the hydroperoxides ROOH in presence of O_2 can lead to development of the multiple chain radiolysis of the oxidative destruction of hyaluronan, as revealed by the increasing loss of viscosity during the irradiation of the aerated aqueous solutions.

Several studies presented the results of chromatographic isolation of several final oligosaccharide products of oxidative hyaluronan destruction [62,74]. The methods IR, UV and NMR spectroscopy, as well as mass-spectrometry, reveal their chemical structure (Figure 5.16).

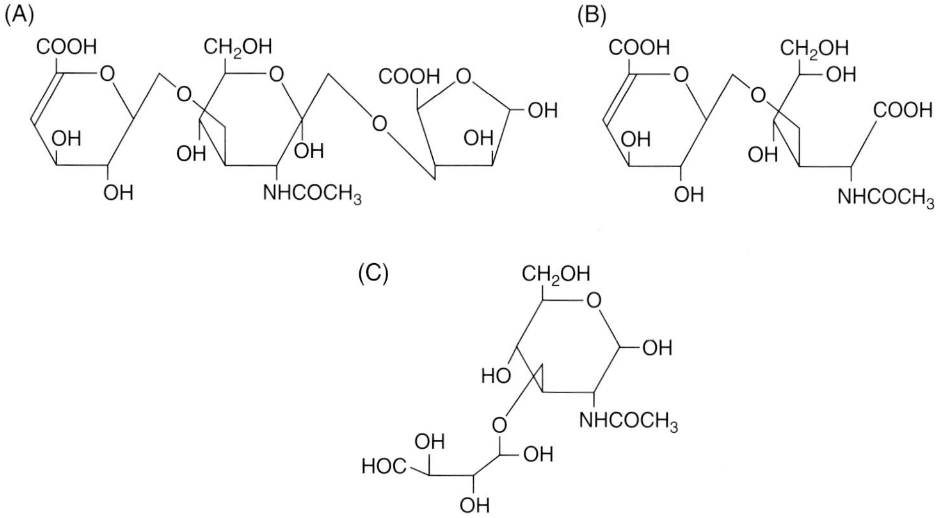

Figure 5.16 *The final products of hyaluronan oxidative destruction,(A) 4,5 unsaturated GlcUA(d1–3)GlcNAc(b1–3)-D-pentauronic acid; (B) 4,5 unsaturated Glc UA(b1–3)-N-acethyl-D- glucosamine acid; (C) L-tri-tetradialdocyl-(b1–3)GlcNAc*

Radiolysis of hyaluronan aqueous solutions could be described as destructive oxidative processes in which hydroxyl radical OH* plays the major role. Therefore, any processes that lead to the formation of the hydroxyl radical, whether in scientific experiment or in nature, will be accompanied by the irreversible destruction of the HA with the formation of low molecular weight fragments of different chemical structures. Understanding of the mechanism of radiolysis of hyaluronan is necessary in order to develop a strategy to protect the polysaccharide from the destructive effects of oxidative radicals.

References

[1] Lowman, A.M., Peppas, N.A. (2000) Molecular analysis of interpolymer complexation in graft copolymer networks. *Polymer*, **41** (1), 73–80.
[2] Wallace, D.G., Rosenblatt, J. (2003) Collagen gel systems for sustained delivery and tissue engineering. *Advanced Drug Delivery Reviews*, **55** (12), 1631–1649.
[3] Gehrke, C.W., Zumwalt, R.W., Walter, A. Aue, W.A. et al. (1971) A search for organics in hydrolysates of lunar fines. *Journal of Chromatography*, **54**, 169–183.
[4] Flory, P.J., Rehner, J. (1943) Statistical Mechanics of cross-linked polymer networks. *Journal of Chemical Physics*, **11** (11), 512–520.
[5] Peppas, N.A., Merrill, E.W. (1977) Crosslinked PVA hydrogels as swollen elastic networks. *Journal of Applied Polymer Science*, **21**, 1763–1770.
[6] Peppas, N.A., Bures, P., Leobandung, W., Ichikawa, H. (2000) Hydrogels in Pharmaceutical Formulations. *European Journal of Pharmaceutics and Biopharmaceutics*, **50**, 27–46.
[7] Lowman, A.M., Peppas, N.A. (1999) Solute transport analysis in pH-responsive, complexing hydrogels of poly(methacrylic acid-g-ethylene glycol). *Journal of Biomaterials Science, Polymer Edition* **10**, 999–1009.
[8] Cruise, G.M., Scharp, D.S., Jeffrey, A. Hubbell, J.A. (1998) Characterization of permeability and network structure of interfacially photopolymerized poly(ethylene glycol) diacrylate hydrogels. *Biomaterials*, **19** (14), 1287–1294.

[9] Canal, T., Peppas, N.A. (1989) Correlation between mesh size and equilibrium degree of swelling of polymeric networks. *Journal of Biomedical Materials Research*, **23**, 1183– 1193.

[10] Hennink, W.E., Van Nostrum, C.F. (2002) Novel crosslinking methods to design hydrogels. *Advanced Drug Delivery Reviews*, **54** (1), 13–36.

[11] West, J.L., Hubbell, J.A. (1995) Photopolymerized hydrogel materials for drug delivery applications. *Reactive Polymers*, **25** (2–3), 139–147.

[12] Nguyen, K.T., West J.L. (2002) Photopolymerizable hydrogels for tissue engineering applications. *Biomaterials*, **23** (22), 4307–4314.

[13] Wang, D.-A., Williams, C.G., Li, Q. et al. (2003) Synthesis and characterization of a novel degradable phosphate-containing hydrogel, *Biomaterials*, **24** (22), 3969–3980.

[14] Drury, J.L., Mooney, D.J. (2003) Hydrogels for tissue engineering: scaffold design variables and applications. *Biomaterials*, **24** (24), 4337–4351.

[15] Andrianov, A.K., Cohen, S., Visscher, K.B. et al. (1993) Controlled release using ionotropic polyphosphazene hydrogels. *Journal of Controlled Release*, **27** (1), 69–77.

[16] Zhang, R., Li, W., Liang, W. et al. (2003), Effect of hydrophobic group in polymer matrix on porosity of organic and carbon aerogels from sol–gel polymerization of phenolic resole and methylolated melamine. *Microporous and Mesoporous Materials*, **62** (1–2), 17–27.

[17] Burger, K., Nad,' G.T., Retei, I. et al. (1990) Associates of deprotonated hyaluronic acid, the method of obtaining the same, pharmaceutical composition containing associates of deprotonated hyaluronic acid and method of obtaining the same (in Russian). Russian Federation Patent RU 2099350, filed Oct. 23, 1990, issued Dec. 20, 1997 Бургер, К., Надь, Г.Т., Ретеи, И. и др. (1990) Ассоциаты депротонированной гиалуроновой кислоты, способ их получения, фармацевтическая композиция, содержащая ассоциаты депротонированной гиалуроновой кислоты, и способ ее получения. Патент РФ № 2099350, filed Oct. 23, 1990, issued Dec. 20, 1997.

[18] Aeschlimann, D., Bulpitt, P. (1998) Functionalized derivatives of hyaluronic acid, formation of hydrogels *in situ* using same, and methods for making and using same. US Patent 6630457 B1, filed Sep 18, 1998, issued Oct. 07, 2003.

[19] Kuo, J.W., Swann, D.A., Prestwich, G.D. (1991) Chemical modification of hyaluronic acid by carbodiimides. *Bioconjugate Chemistry.* **2** (4), 232–241.

[20] Aeschlimann, D., Bulpitt, P. (2003) Functionalized derivatives of hyaluronic acid, formation of hydrogels *in situ* using same, and methods for making and using same. US Patent 7196180 B2. filed Oct. 06, 2003, and issued Mar. 27, 2007.

[21] Tomihata, K., Ikada, Y. (1997) Crosslinking of hyaluronic acid with water-soluble carbodiimide. *Journal of Biomedical Materials Research*, **37** (2), 243 – 251.

[22] Lu, P.L., Lai, J.Y., Ma, D.H., Hsiue, G.H. (2008) Carbodiimide cross-linked hyaluronic acid hydrogels as cell sheet delivery vehicles: characterization and interaction with corneal endothelial cells. *Journal of Biomaterials Science Polymer Edition*, **19** (1), 1–18.

[23] Young, J.J., Cheng, K.M., Tsou, T.L. et al. (2004) Preparation of cross-linked hyaluronic acid film using 2-chloro-1-methylpyridinium iodide or water-soluble 1-ethyl-(3,3-dimethylaminopropyl)carbodiimide. *Journal of Biomaterials Science Polymer Edition*, **15** (6), 767–780.

[24] Sannino, A., Pappadà, S., Madaghiele, M. et al. (2005) Crosslinking of cellulose derivatives and hyaluronic acid with water-soluble carbodiimide, *Polymer*, **46** (25), 11206–11212.

[25] Kuo, J.W., Swann, D.A., Prestwich, G.D. (2000) Water-insoluble derivatives of hyaluronic acid crosslinked with a biscarbodiimide. US Patent 6537979, filed Mar 31, 2000, issued Mar 25, 2003.

[26] Lai, J.Y. (2012) Solvent composition is critical for carbodiimide cross-linking of hyaluronic acid as an ophthalmic biomaterial. *Materials* **5** (10), 1986–2002.

[27] Renier, D., Crescenzi, V., Francescangeli, A. (2001) Crosslinked derivatives of hyaluronic acid. US Patent 7125860 B1, filed Aug. 31, 2001, and issued Oct. 24, 2006.

[28] Bulpitt, P., Aeschlimann, D. (1999) New strategy for chemical modification of hyaluronic acid: preparation of functionalized derivatives and their use in the formation of novel biocompatible hydrogels. *Journal of Biomedical Materials Research*, **47** (2), 152–169.

[29] Vercruysse, K.P., Marecak, D.M., Marecek J.F., Prestwich, G.D. (1997) Synthesis and *in vitro* degradation of new polyvalent hydrazide cross-linked hydrogels of hyaluronic acid. *Bioconjugate Chemistry.* **8** (5), 686–694.

[30] Alexander, C., Fraser, J.E., Zhao, X. (2000) Process for cross-linking hyaluronic acid to polymers, WO Patent 2000046252 A1, filed Feb. 3, 2000, and issued Aug. 10, 2000.

[31] Tomihata, K., Ikada, Y. (1997) Preparation of crosslinked HA films of low water content. *Biomaterials*, **18** (3), 189–195.

[32] Hamilton, R., Fox, E.M., Acharya, R.A., Walts, A.E. (1987) Water insoluble derivatives of hyaluronic acid. US Patent 4937270, filed Sep. 18, 1987, and issued Jun. 26, 1990.

[33] Balazs, E., Leshchiner, A. (1984) Cross-linked gels of hyaluronic acid and products containing such gels. US Patent 4582865, filed Dec. 06, 1984, and issued Apr. 15, 1986.

[34] Della Valle, F., Romeo, A. (1986) Esters of hyaluronic acid. US Patent 4851521, filed Jul. 02, 1986, issued Jul. 25, 1989.

[35] Miyamoto, K., Waki, M. (1996) Photocured cross-linked-hyaluronic acid gel and method of preparation thereof. US Patent US6031017, filed Nov. 14, 1996, and issued Feb. 29, 2000.

[36] Laurent, T.C. (1967) Determination of the structure of agarose gels by gel chromatography. *Biohimica et Biophisica Acta*, **136**, 199–205.

[37] De Belder, A.N., Malson, T. (1989) Method of preventing adhesion between body tissues, means for preventing such adhesion, and process for producing said means. US Patent 4886787, filed Jan. 23, 1986, issued Dec. 12, 1989.

[38] Nishi, C. (1995) *In vitro* evaluation of cytotoxicity of diepoxy compounds used for biomaterial modification, *Journal of Biomedical Materials Research*, **29** (7), 829– 834.

[39] Sakurai, K., Ueno, Y., Okuyama, T. (1985) Crosslinked hyaluronic acid and its use. Patent US 4716224, filed May 2, 1985, issued Dec. 29, 1987.

[40] De Belder, A., Malson, T. (1985). Gel for preventing adhesion between body tissues and process for its production. WO Patent 1986000912 A1, filed Jul. 16, 1985, issued Feb. 13, 1986.

[41] Schante, C.E., Zuber, G., Herlin, C., Vandamme, T.F. (2011) Chemical modifications of hyaluronic acid for the synthesis of derivatives for a broad range of biomedical applications. *Carbohydrate Polymers*, **85**, 469–489.

[42] Matsuda, T., Moghaddam, M., Sakurai, K. (1993) Photocurable glycosaminoglycan derivatives, crosslinked glycosaminoglycans and method of production thereof. US Patent 5462976, filed Feb. 05, 1993, issued Oct. 31, 1995.

[43] Martens, P., Anseth, K.S. (2000) Characterization of hydrogels formed from acrylate modified poly(vinyl alcohol) macromers, *Polymer*, **41** (21), 7715–7722.

[44] Baier Leach, J, Bivens, K.A., Patrick, C.W. Jr., Schmidt, C.E. (2003) Photocrosslinked hyaluronic acid hydrogels. Natural, biodegradable tissue engineering scaffolds. *Biotechnology & Bioengineering*, **82**, 578–589.

[45] Shu, X.Z., Prestwich, G.D. (2004) therapeutic biomaterials from chemically modified hyaluronan, in *Chemistry and Biology of Hyaluronan* (eds H.G. Garg, C.A. Hales), Elsevier, Amsterdam, pp. 475–504.

[46] Raeber, G.P., Lutolf, M.P., Hubbell, J.A. (2005) Molecularly engineered PEG hydrogels: A novel model system for proteolytically mediated cell migration. *Biophysical Journal*, **89** (2), 1374–1388.

[47] Khabarov, V.N., Selyanin, M.A., Zelenetsky, A.N. (2008) Solid-state modification of hyaluronic acid for applications in esthetic medicine (in Russian). *Vestnik Estetichskoi Mediciny*, **7** (3), 18–24. Хабаров, В.Н., Селянин, М.А., Зеленецкий, А.Н. (2008) Твердотельная модификация гиалуроновой кислоты для целей эстетической медицины, *Вестник эстетической медицины*, **7** (3), 18–24.

[48] Bridgman, P.W. (1935) Effects of high shearing stress combined with high hydrostatic pressure. *Physical. Review*, **48**, 825–847.

[49] Enikopolov, N.S. (1991) Solid phase chemical reactions and new technologies, *Russian Chemical Reviews*, **60** (3), 283–287.

[50] Akopova, T.A., Zhorin, V.A., Volkov, V.P. et al. (2007) The method of obtaining cross-linked salts of hyaluronic acid (in Russian). Russian Federation Patent RU 2366665, filed Dec. 03, 2007, issued Sep. 10, 2009. Акопова, Т.А., Жорин, В.А., Волков В.П. и др (2007) Способ получения сшитых солей гиалуроновой кислоты. Патент № RU 2366665 filed Dec. 03, 2007, issued Sep. 10, 2009.

[51] Khabarov, V.N., Selyanin, M.A., Michailova, N.P., Zelenetsky, A.N. (2009) Bioactive compositions comprising hyaluronic acid modified in solid-phase (in Russian). *Vestnik Estetichskoi Mediciny*, **8** (1), 49–53 Хабаров, В.Н., Селянин, М.А., Михайлова, Н.П., Зеленецкий, А.Н. (2009) Биоактивные композиции на основе твердофазно модифицированной гиалуроновой кислоты. *Вестник эстетической медицины*, **8** (1), 49–53.

[52] Khabarov, V.N., Selyanin, M.A., Zelenetsky, A.N. (2008) Perspectives of development of new preparations for biorevitalization (in Russian). *Vestnik Estetichskoi Mediciny*, **7** (4), 40–45. Хабаров, В.Н., Селянин, М.А., Зеленецкий, А.Н. (2008) Перспективы создания новых препаратов для биоревитализации. *Вестник эстетической медицины*, **7** (4), 40–45.

[53] Khabarov, V.N., Zelenetsky, A.N. (2008) Nanotechnological reticualtion of hyaluronic acid (in Russian). *Kosmetik International*, **2**, 8–15. Хабаров В.Н., Зеленецкий А.Н. (2008) Нанотехнологическая ретикуляция гиалуроновой кислоты. *Kosmetik International*, **2**, 8–15.

[54] Lal, M. (1985) Radiation induced depolymerization of HA in aqueous solution at pH 7.4. *Journal of Nuclear and Radiochemical Sciences*, **92** (1), 105–112.

[55] Alassat, S., Hawkins, C.L., Parsous, B.J. (1999) Identification of radicals from HA using EPR spectroscopy. *Carbohydrate Polymers*, **38** (1), 17–22.

[56] Garg, H.G., Hales, C.A. (eds) (2004) *Chemistry and Biology of Hyaluronan.* Elsevier, Amsterdam.

[57] McNeil, J.D., Wiebkin, O.W., Betts, W.H. (1985) Depolymerisation of HA after exposure to oxygen derived free radicals. *Annals of the Rheumatic Diseases*, **44** (11), 780–789.

[58] Sonntag, C.V., Bothe, E. (1995) Pulse radiolysis in model studies toward radiation processing. *Radiation Physics and Chemistry*, **46** (4), 527–532.

[59] Myint, P. (1987) The reactivity of various free radicals with HA: pulse radiolysis studies. *Biochimica et Biophysica Acta*, **925**, 194–202.

[60] Srinivas, A., Ramamurthi, A. (2007) Effect of gamma-irradiation on physical and biologic properties of HA. *Tissue Engineering*, **13** (3), 447–459.

[61] Visco, A.M., Campo, N. Torrisini, L. (2008) Electron beam irradiated: degrading action of air on HA. *Biomedical Materials and Engineering*, **18** (3), 137–148.

[62] Kim, J.K., Srinivasan, P., Kim, J.H., Choi, J., Park H.J. (2008) Structural and antioxidant properties of gamma-irradiation HA. *Food Chemistry*, **109** (4), 763–770.

[63] Rice-Evans, C.A., Burdon, R.H. (1994) *Free Radical Damage and its Control.* Elsevier Science, Amsterdam.

[64] Soltes, L., Kogan, G., Stankovska, M. (2007) Degradation of HA and characterization of fragments. *Biomacromolecules*, **8** (9), 2697–2705.

[65] Pikaev, A.K., Kabakchi, S.A., Makarov, I.E., Ershov, B.G. (1980) *Pulse Radiolysis and its Applications* (in Russian), Atomizsdat, Moscow, p. 268 bПикаев, А.К., Кабакчи, С.А., Макаров, И.Е. (1980) *Импульсный радиолиз и его применение.* Атомиздат. Москва, С.268.

[66] Sharpatyi, V.A. (2006) *Radiation Chemistry of Biopolymers.* CRC Press, Taylor & Francis Group, Boca Raton.

[67] Balazs, E.A., Davies, J.V., Phillips, G.O. (1967) Transient intermediates in the radiolysis of hyaluronic acid. *Radiation Research*, **31** (2), 243–255.

[68] Khabarov, V.N., Kozlov, L.L., Panchenkov G.M. (1980) Gamma-radiolysis of aqueous solutions of methyl orange and chrysoidine (in Russian). *Khimiya Vysokikh Energii*, **14** (5), 406–408. Хабаров, В.Н., Козлов, Л.Л., Панченков, Г.М. (1980) Гамма-радиолиз водных растворов метилоранжа и хризоидина. *Химия высоких энергий*, **14** (5), 406–408.

[69] Khabarov, V.N., Kozlov, L.L., Panchenkov G.M. (1981) Kinetic dependences of radiation-chemical reduction of azo dyes in aqueous solutions (in Russian) *Zhurnal Fizicheskoi Khimii*, **55** (12), 3072–3075. Хабаров, В.Н., Козлов, Л.Л., Панченков, Г.М. (1981) Кинетические закономерности радиационно-химического восстановления азокрасителей в водных растворах, **55** (12), 3072–3075.

[70] Khabarov, V.N., Kozlov, L.L., Panchenkov G.M. (1981) Mechanism and kinetics of radiation discoloration of azo dyes (in Russian) *Khimiya Vysokikh Energii*, **15** (3), 218–221. Хабаров, В.Н., Козлов, Л.Л., Панченков, Г.М. (1981) Механизм и кинетика радиационного обесцвечивания азокрасителей. *Химия высоких энергий*, **15** (3), 218–221.

[71] Khabarov, V.N., Selyanin, M.A., Michailova, N.P. (2011) Evaluation of antioxidant efficacy of some compounds used in estetic medicine (in Russian). *Vestnik estetichskoi mediciny*, **10** (1), 52–56. Хабаров, В.Н., Селянин, М.А., Михайлова, Н.П. (2011) Оценка антиоксидантной эффективности некоторых соединений, применяемых в эстетической медицине. *Вестник эстетической медицины*, **10** (1) 52–56.

[72] Jooyanden, F., Moore, V.S., Morgan, R.E., Phillips, G.O. (1971) Chemical effects of gamma-irradiation of aqueous solution of heparin and keratin sulfate. *Radiation Research*, **45** (3), 455–461.

[73] Issels, R.D., Fink, R.M., Lengbelder, E. (1986) Effect of hyperthermic conditions on the reactivity of oxygen radicals. *Free Radical Research*, **2** (1–2), 7–18.

[74] Kuo, J.W. (2005) *Practical Aspects of Hyaluronan Based Medical Products*. CRC Press, Taylor & Francis Group. Boca Raton.

6

Medical Applications of Hyaluronan

6.1 Hyaluronan and Aesthetic Medicine

It is difficult to imagine cosmetology and the field of modern aesthetic medicine without hyaluronan. A glance at the worldwide periodical literature would reveal that practically every issue includes several articles more or less connected with HA. Physician-cosmetologists were initially interested in hyaluronan because of its unique ability to retain large amounts of water. This property has been maximized for cosmetic products and moisturising masks. From the mid-nineteenth to the twentieth century, chemically stabilized (cross-linked) HA became used as intradermal micro-implants (fillers) for the contour correction of the involuntary changed skin [1].

The development of injection cosmetology is tightly connected with methods of contour plastic, mesotherapy and bio-revitalization, for which an enormous number of products containing non-stabilized or partially modified HA have appeared. The cosmetologist's arsenal has recently been enriched by a new class of products: non-medicated macromolecular therapeutic agents called bio-repairs in which biologically active compounds are chemically immobilized on the hyaluronan macromolecules. With the aid of these sustained action formulations, it is now possible to realize target delivery of vitamins, amino acids and oligopeptides to the specific cells for purposes of skin homeostasis stabilization and prevention of the age-related changes.

6.1.1 Intradermal Hyaluronan-Based Microimplants

The development history of the first intra-dermal hyaluronan-based microimplants warrants brief review. Since the early 1980s, HA microimplant injections have been commonly and widely used for correction of depressive cosmetic defects. In 1984–1988, the preparation

Hyaluronic Acid: Preparation, Properties, Application in Biology and Medicine, First Edition.
Mikhail A. Selyanin, Petr Ya. Boykov and Vladimir N. Khabarov.
© 2015 John Wiley & Sons, Ltd. Published 2015 by John Wiley & Sons, Ltd.

Hylan B, which consists of hyaluronan chemically cross-linked with divinylsulfone (DVS), was developed in the USA. The first mention on the cross-linking of HA by epoxy compounds dates back to the early 1960s, but the first pre-clinical and clinical tests for correction of soft tissues were carried out for Hylan B from 1990–1994. Around the same time, in 1993, on the basis of HA cross-linked with 1,4-butanediol diglycidyl ether (BDDE), the product Restylane was approved in Europe and was positioned as an intra-dermal filler. Today, the number of similar products that include reticulated hyaluronan and are produced in different countries has reached several dozen. HA, as acknowledged by practically all specialists in aesthetical medicine, is almost ideal material for the implantation. According to the report of the American society of aesthetical and plastic surgery (ASAPS), the injection of hyaluronan occupies second place among all other nonsurgical methods of the skin rejuvenation with 1 313 038 injections in the USA in 2009 alone, which is second in use only to the injections of botulinum toxin type A (2 557 068 injections in the USA in 2009). By accomplishing the principal task of providing additional volume in the sites of tissue deficit (e.g. wrinkle, fold, lipodystrophy zones, small skin defects due to 'less-tissue' effects after surgical interventions and injuries), hyaluronan implants remain in the skin for a long time, do not have side effects, do not require allergic tests and application is simple and painless during injection.

As the technology of cross-linking was optimized and perfected, success was achieved in designing HA microimplants. In particular, some contemporary hyaluronan-based fillers are characterized by their considerably long times of active correction, that is, up to 15–18 months. Despite the fact that varying HA-based dermal fillers have much in common, they all, nevertheless, differ in consistency, concentration, degree of reticulation and chemistry of bi-functional cross-linking agents used; differentiations that manifest in different degrees of correcting effect.

In the next section, the properties of hyaluronan microimplants are discussed from the contemporary viewpoint of physical chemistry of gel-like polymeric materials.

6.1.2 Cross-Linking of Hyaluronan into a Three-Dimensional Network

In the process of preparing material that can be used as a HA-based dermal implant, the main step concerns the chemical modification of the polysaccharide macromolecule. This modification occurs through formation of the three-dimensional structure as a result of a reaction with a bi-functional cross-linking agent. The compounds used for such a chemical modification are discussed in detail in Section 5.2. These bi-functional reagents contain active groups and are capable of reacting with the functional groups of the polysaccharide macromolecule.

The main 'targets' for chemical modification in the hyaluronan molecule are shown in Figure 6.1. Currently, the most popular cross-linking agent used in the production of HA-based fillers is the epoxide derivative 1,4-butanediol diglycidyl ether (Figure 6.2), and the process of production itself is analogous to the reaction that takes place during the hardening of epoxy resin.

In the USA, the FDA permits both the use of divynilsulfone (DVS) (Figure 6.3) and BDDE as cross-linking agents.

In cross-linking HA with bi-functional reagents like BDDE or DVS, first an aqueous solution of hyaluronan is prepared, into which the given quantity of the cross-linking agent

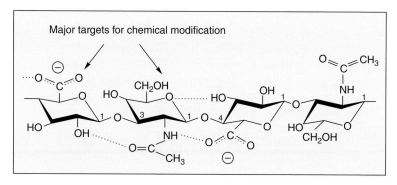

Figure 6.1   *Main chemical groups of HA that react with bi-functional cross-linking agents*

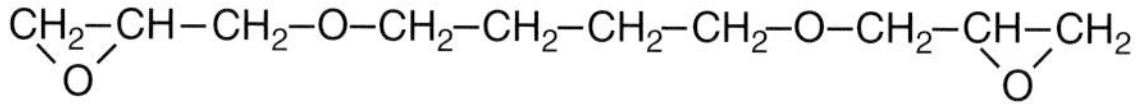

$$CH_2-CH-CH_2-O-CH_2-CH_2-CH_2-CH_2-O-CH_2-CH-CH_2$$

Figure 6.2   *Chemical structure of the BDDE molecule*

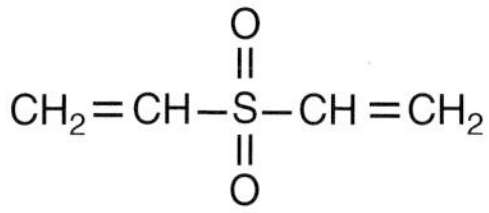

$$CH_2=CH-S-CH=CH_2$$

Figure 6.3   *Chemical structure of the DVS molecule*

is added. The reaction is carried out at a different pH and, depending on the reaction conditions, either simple or complex ester bonds are formed in the cross-linked HA chains (see Sections 5.2.3 and 5.2.5 in Chapter 5). If one assumes that all macromolecular chains of the polysaccharide are interconnected through three-dimensional bonds, then an entire volume of polymer can be considered to be one gigantic macromolecule. In this case, one can consider a monophase cross-linked hyaluronan gel. A similar situation is realized in the synthesis of polyacrylamide gel [2] or radiation-induced cross-linking of polyoxyethylene [3]. It is, however, now well-known [4] that structured biopolymers like HA do not form regular three-dimensional networks with the sequential alternation of cross-links, as this is generally a thermodynamically unfavourable process. The block of this three-dimensionally structured biopolymer is generally built from the cross-linked macromolecules of hyaluronan of globular structure (for more detail see Chapter 4). Therefore, in creating a high degree of intermolecular cross-linking, as this determines the time of dermal filler residence in the place of implantation, it is impossible to obtain uniform monophase gel when the process is carried out in the liquid phase. Recent studies [4] show that cross-linked HA is characterized by significant structural heterogeneity, that is, instead of the gigantic macromolecule the structure of gel is composed of discrete particles of microscopic sizes that represent three-dimensional micro-heterogeneous blocks connected with the non-reticulated hyaluronan macromolecules via hydrogen bonds (Figure 6.4).

The scanning electron micrograph of HA based cross-linked hydrogel nanoparticles is presented in Figure 6.5.

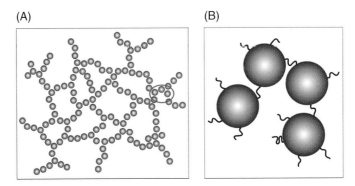

Figure 6.4   *Schematic representation of the intra-particle cross-linking (A) and inter-particle cross-linking (B) [4]*

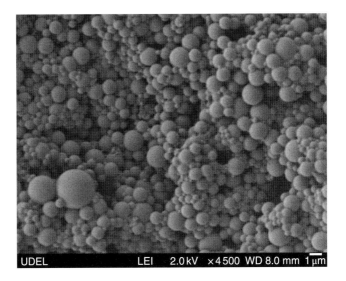

Figure 6.5   *Scanning electron micrograph of HA cross-linked hydrogel particles [4]. Reprinted with permission from [4]. Copyright © 2009, American Chemical Society*

Further polymerization results in an increase in grain size following their rapprochement and interaction with each other by means of the cross-linking agent. The density of cross-links in such zones is high, which leads to the formation of a rigid micro-heterogeneous hydrophobic structure consisting of the highly cross-linked extended elements (grains, bodies) unevenly distributed in the hydrogel. This, in turn, leads to an increase in viscosity, loss of fluidity of the polysaccharide solution and corresponds to the gelation point. On the basis the statistical theory developed for cross-linked polymers [5], it is possible to calculate the degree of the cross-linking defined as a quantity of cross-links per 100 disaccharide units of HA using the sol-gel method of analysis. In studies [6,7] the authors carried out comparative analyses of the structure of the cross-linked HA hydrogels obtained using the traditional liquid-phase method with that synthesized using the technology of solid-phase modification (the method of solid-phase synthesis is described in the Section 5.2.7 in Chapter 5). It was established that different methods of HA cross-linking result in a substantial change in the spatial structure of the cross-linked gels. When using the traditional liquid-phase technology of HA reticulation with a relatively high degree of cross-linking, it is impossible to achieve uniform distribution of cross-links in the sample. With the overall reticulation degree of nine cross-links per 100 disaccharide units in HA, the gel comprised the regions of the densely cross-linked structures along with the zone where macromolecules did not undergo chemical modification. The problem with such a method of polymer system reticulation is associated with the heterogeneous molecular-weight distribution of the raw hyaluronan material. Practically in all HA materials, independently of the method of their production, there is always a wide collection of polysaccharide fractions of different molecular weight (the molecular-mass distribution is discussed in more details in the next section). According to statistical laws, upon cross-linking of such polydisperse systems under mild conditions of aqueous solution, the high-molecular weight component of hyaluronan is cross-linked considerably more rapidly in comparison with the low-molecular weight fractions. Hence, the heterogeneity of the resultant gel occurs due to concentration of cross-links inside the macromolecular coil. This problem can be solved to a certain extent by using the technology of the solid-phase modification of polysaccharides (the procedure of so-called solid-phase reactive displacement or solid-state reactive blending (SSRB) [8].

With this method of chemical process initiation, the segments of the HA macromolecule are oriented along the plastic flow plane of the polymer (such regions are approximately 3–5 nm) and the molecules of the cross-linking agent are located in the same regions. In this case, the nano-dimensional regions where the cross-linking reaction occurs are quasi-uniformly distributed along the entire plane of the applied pressure and, correspondingly, the reaction occurs evenly throughout the entire thickness of the sample with 100% degree of BDDE transformation. The degree of HA reticulation is determined by the quantity of the cross-linking component taken for synthesis. The use of such solid-phase modification allowed authors [6,7,9] to obtain HA gel with the degree of reticulation of 16 cross-links per 100 disaccharide units of HA. This was the maximal achievable degree of cross-linking at which the resultant material did not lose its viscoelastic properties required for use as an injectable implant.

It should be noted that the material obtained using the aforementioned application of innovative solid-phase technology can be attributed to the monophase (homogeneous) fillers. However, this monophase material is completely biodegraded in the site of injection after 4–6 months. Prolongation of the retention time of this intradermal implant by increasing the degree of HA reticulation while preserving monophasic structure of the material appeared to be impossible. The value of 16 cross-links per 100 disaccharide units is apparently the maximal cross-linking degree, since its further increase led to a drop in the material's elasticity.

Let us discuss this in further detail. 'Elasticity' is the ability of material to restore its initial form after removal of stress exerted by external forces. When the form is restored completely, the material is called 'elastic.' The polymeric materials that possess elastic properties can be divided into two groups. The first group includes materials that manifest very strong resistance to a change in the form and are reversibly deformable to an extent. The second group includes polymers that easily change their form and are capable to be reversibly deformed to many hundred percent. In many respects, these properties can be described by two basic parameters frequently used in physical chemistry of polymers [4], namely shear modulus (modulus of rigidity) G and modulus of elasticity E. The modulus of rigidity and modulus of elasticity relates as follows:

$$2G_T(1+m) = E \tag{6.1}$$

where m is a Poisson's ratio (dimensionless quantity; usually for polysaccharides $m=0.5$).

In the statistical theory of elasticity for spatially structured polymers, the modulus of elasticity is a function of the number of effective cross-links:

$$E = \frac{3RTd}{M_c} \tag{6.2}$$

where R is universal gas constant, T is a temperature, d is polymer density, M_c is a molecular weight of polymer chain fragment in between two cross-links.

The network density or the number of cross-links is determined either as Mc or molar concentration of the chain fragments between two cross-links (v) in the unit of volume (V):

$$N = \frac{v}{V} = \frac{d}{M_c} \tag{6.3}$$

where N is a network density or number of effective cross-links.

These relationships used in the theory of elasticity and developed for ideal networks make it possible to determine the degree of polysaccharide reticulation and establish the maximal degree of cross-linking in the hyaluronan gel, after which its injection into derma becomes technically impossible. Specifically, this determines comparatively short (up to half a year) times of the correction of the skin volume changes by intradermal administration of the cross-linked HA-based monophase (homogeneous) fillers. The solution to the problem of increasing the duration of correction lies in the field of bi-phase (heterogeneous) micro-implant administration.

One of the stages in the process of preparing dermal HA fillers consists in breaking up densely cross-linked gel mass into small particles (the degree of reticulation of the hyaluronan macromolecules in such gel particles can be as high as 30–40 cross-links per 100 units [10]), followed by sifting the crushed gel mass through several lattice filters or sieves. In the next step, these densely cross-linked gel particles are suspended in the solutions of non-stabilized HA, making the size of the particles to vary in narrow interval, as determined for products manufactured by different suppliers (Figure 6.6).

Such hyaluronan microimplants are characterized by the following physico-chemical properties:

- They are bi-phasic, that is, there is a phase interface between the components of system.
- There is an affinity between the components through the forming part of the material, and non-stabilized and densely cross-linked gel particles, which leads to the formation of a thermodynamically and aggregation-stable material.

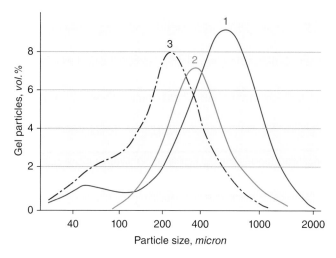

Figure 6.6 *Particle size distribution in the HA-based products Restylane (1), Perlane (2) and Hyalaform (3) [11]. Reproduced from [11] with permission from Taylor & Francis*

- Specific distribution of gel particles (beads) according to size. Contemporary two-phase fillers are characterized by relatively uniform size of the cross-linked gel beads (Figure 6.5). The exception is the fillers obtained by mechanical homogenization of the gel masses that results in uniform consistency of the end product, but leads to a broader particle size distribution.

Fundamental differences between bi-phase and mono-phase intradermal HA fillers are as follows:

1. Bi-phasic HA gel contains densely cross-linked polysaccharide particles with a large number of chemical bonds. This leads to a reduced ability of the biopolymer to absorb water; because of the dense three-dimensional network, cross-linked HA almost completely loses its ability to swell. In the mono-phase systems the number of cross-links is comparatively small (i.e. the chain segments between chemical cross-links are relatively large and water molecules can penetrate the polymeric phase to induce limited swelling). Since the initial stage of cross-linked hyaluronan structure degradation predominantly involves non-enzymatic hydrolysis of the chemical bonds between the cross-linked moieties and HA macromolecule, this process resumes at higher rates in the mono-phase material.
2. Bi-phasic gels contain non-stabilized HA that plays the role of plasticizer of a type of exerting bio-revitalizing effect.
3. Bi-phasic fillers contain HA with a higher degree of reticulation compared to mono-phase systems, hence the content of chemical modifier BDDE or DVS bound to HA is 2–3 times higher in the bi-phasic gels. Upon biodegradation of microimplants, there is a probability of free molecule formation of ethylene-bis-1,2-propanediol $HOCH_2CH(OH)$ $CH_2O(CH_2)_4OCH_2CH(OH)CH_2OH$ or thionyl-bis-ethanol $HOCH_2CH_2SO_2CH_2CH_2OH$. Although these compounds are known to be non-cytotoxic, their metabolism in human bodies is not sufficiently studied to exclude possibility of development of undesired side-effects post implantation.

In summary, the currently marketed dermal HA-based microimplants are close in their characteristics to 'ideal' filler due to bio-inertness, simplicity of administration and duration of correction action in the organism. However, one cannot ignore possible potential risk associated with prolonged application of such materials. The chemical structure of modified HA now comprises -CH_2- groups from the cross-linking agent, which is unusual for native hyaluronan macromolecules. Under the action of UV irradiation, free radicals (mainly hydroxy-radicals) are formed in the skin, which, in the presence of oxygen, react with -CH_2- through the free-radical chain oxidation process [12]. As a result of such oxidizing reactions, cross-linked HA may exhibit pro-oxidant properties, that is, become a source of free-radicals that attack and damage the structure of the collagen and dermal proteins in the sites of implantation. Therefore, the main efforts of the scientist working with HA materials focus on the development of the physico-chemical methods of HA modification that do not require the use of bi-functional reagents.

6.1.3 Hyaluronic Acid in Injection Cosmetology (Biorevitalization)

The content of hyaluronan in the human skin is variable. There are insignificant seasonal variations in the dermal content of HA: in summer the level of hyaluronan is somewhat low compared to the winter. This occurs because of a higher rate of HA degradation under the action of UV radiation [13]. The most significant effect is the age-related decrease in HA concentration. The study [11] analyses the numerous data on the measurements of the hyaluronan content in different age groups of patients. Starting from the age of 60 years, the concentration of HA in derma rapidly decreases. Therefore, intradermal injections of native HA represent a quite natural method of biopolymer replenishing. This injection method in aesthetical medicine is called biorevitalization. It is believed that the term 'biorevitalization' was proposed in 2001 by Italian researcher Di Pietro, who originally defined it as a method of the intradermal injections of unmodified hyaluronan that makes it possible to restore the physiological medium and normalize the metabolic processes in derma [14].

Consequently, biorevitalization can primarily be considered as a method of extracellular matrix restoration, which in turn determines the optimal algorithm of skin cell bioactivity. Administration of native HA leads to an improvement in the proliferating activity of fibroblasts, stimulation of the cell-mediated collagen and elastin synthesis, simulation of the differentiation of fibroblasts into fibroblasts and activation of angiogenesis. Ten years of experience with native HA application in aesthetical medicine has allowed for the fundamental characteristics of biorevitalizers to be experimentally determined. First, the molecular weight of hyaluronan and the concentration of active material ranges from 0.5 to 1.8 mass in percentage (the optimal weight is considered to be approximately 1 MDa). The range of working HA concentrations in biorevitalizers is determined by ease of administration and by the absence of the undesirable side effects in the form of dermal oedema. The molecular weight of hyaluronan is an extremely important parameter since the macromolecules of different weights influence cell behaviour in different ways.

It is assumed [15] that molecular weight plays the most important role in the mechanisms of physiological regulation. The relatively low-molecular weight fraction of HA (i.e. less than 10 000 Da) exhibits inflammatory action and induces angiogenesis (proliferation of the blood and lymphatic vessels). Hyaluronan with the molecular weight of 50 000–100 000 Da stimulates cellular proliferation and activates the migration of cells. The fraction of HA with the molecular weight of more than 500000 suppresses angiogenesis, inhibits cellular proliferation, and blocks secretion of interleukin 1b and prostaglandin E2 (mediators of inflammation). In the study [16], it was experimentally shown that the effectiveness of fibroblast protection from the cytotoxic action of free radicals correlatively increases with the size of the HA macromolecules. The most effective in this respect is hyaluronan with the molecular weight of about 1 MDa. The aforementioned study data demonstrates that macromolecules of HA with different lengths of polysaccharide chains possess a unique ability to exert completely opposite effects at the molecular and cellular level. A convincing explanation of this phenomenon is yet to be found, as are the yet to be discovered regulation mechanisms besides the unique 'hyaluronan–receptor' system, wherein the signal transfer changes depending on the molecular dimension of the ligand. It thus follows that the molecular weight of HA in biorevitalizers should be strictly controlled to include only relatively narrow fractions with a defined molecular weight. The molecular weight of hyaluronan in such formulations is therefore of paramount importance. A discussion of the types of molecular weight used for characterization of HA-based biorevitalizers is provided in the following section.

6.1.4 Molecular Weight of Hyaluronan in Biorevitalization Products

It is well known that existing data on the molecular weight of HA should be evaluated with caution since obtaining a clean and intact polysaccharide with the high degree of polymerization and 100% yield is practically an impossible task (see Section 6.2). Any sample of polysaccharide absolutely not homogeneous because it is composed of a mixture of polymer homologues – chains of which contain an extra repeating unit of polymer. Practically all samples of HA, whether isolates from raw animal material or a biopolymer of bacterial origin, contain a mixture of different-length macromolecules that are more or less uniform in mass. Therefore, molecular weight of the polymer is described by some average value. Depending on the method used for determining and calculating the average value, the average weight value of samples may appear differently. Usually, the calculated value is the number-averaged Mn and weight-averaged Mw molecular weights:

$$M_n = \frac{\sum N_i M_i}{\sum N_i} \tag{6.4}$$

$$M_w = \frac{\sum N_i M_i^2}{\sum N_i M_i} \tag{6.5}$$

where N_i is the number of macromolecules of weight M_i.

Let us consider a simple numerical example [4]. If there is a mixture of macromolecules of different mass, say, 100 molecules of weight 1000 Da, 200 molecules of 10000 Da, and 200 more molecules of 100000 Da, then the molecular weight values calculated according to formula (6.4) and (6.5) would be: $M_n = 44000$ Da and $M_w = 91000$ Da, respectively. It is clear from the given example that the numerical average of M_n for the polymodal distribution (i.e. presence of fractions of different macromolecule lengths) does not coincide with the weight-averaged M_w. In certain cases [4], the difference in the values of molecular weight determined by different methods for one and the same polymer sample may reach five and more times. For example, HA isolated from the culture of *Streptococus equi* (produced in Japan) with a declared molecular weight of 1.6 MDa appears to have a weight of 480000 Da when analysed by means of light-scattering technique. Thus, the more heterogeneous, in other words more 'polymolecular,' the initial polysaccharide is, the bigger the difference in the average values of molecular weight obtained by different methods. The polydispersity is measured by the ratio M_w/M_n and is used to characterize the molecular weight distribution of the initial unfractionated polysaccharide. This ratio is a fundamental characteristic that is especially important for the medical preparations containing hyaluronan. From the viewpoint of molecular weight distribution, the biopolymer should ideally have monomolecular distribution for which the ratio M_w/M_n approaches unity.

Both M_n and M_w of polysaccharides are determined using physical methods of osmometry, light scattering, diffusion, ultracentrifugation, and viscometry. Molecular weight is most frequently determined by measuring the viscosity of HA solutions. Thanks to the simplicity of the equipment used, by this procedure the average viscometric molecular weight differs from the average of M_w by not more than 20%. The viscometric method is based on the measurement of the reduced viscosity (as a function of polymer concentration), followed by determination of the intrinsic viscosity at zero dilution.

This method has limitations, the first of which is associated with the polydispersity of the sample. It is only possible to establish the direct experimental correlation between the intrinsic viscosity and average molecular weight for polymers that are more or less uniform with respect to molecular weight. In other words, only a relatively narrow fraction of a polymer can be tested by viscometry since even insignificant polydispersity strongly affects the results of intrinsic viscosity determination. Secondly, molecular weight determination by viscometry for polyelectrolytes like polyanionic HA is associated with difficulties. In particular, the presence of the ionized groups (in the case of HA these are carboxyl groups of glucuronic acid) strongly influences the viscosity of the dilute solutions because of mutual repulsion of macromolecules, which leads to a significant increase in viscosity upon dilution. As a consequence, the correlation between reduced viscosity and polymer concentration deviates from a linear one to a more cumbersome extrapolation of reduced viscosity to zero concentration as well as determination of intrinsic viscosity. Therefore, the viscometry method is recommended mostly for determining the relative change in the macromolecule weight in different processes, for example, polymer destruction. This discussion of viscometry as a method for molecular weight determination is provided in detail because most of the HA certificates of analysis on the market states viscometry-based values of molecular weight.

Table 6.1 cites experimental data on the determination of fraction composition and molecular weight for HA from different suppliers [17].

Table 6.1 summarizes the experimentally obtained maximum and minimum molecular weight values in the studied HA materials. The wide range of the average molecular weight values is because of strong heterogeneity in the molecular- weight distribution of macromolecules. The content of low molecular weight fractions reported in Table 6.1 for studied commercial HA was determined by means of gel filtration on sephadex G-50 (fractions of less than 10000 Da) and G-50 (fractions with the weight of less than 150000 Da). The data in Table 6.1 show that the content of these fractions reaches 26% in some HA samples.

Table 6.1 Characteristics of the commercial hyaluronic acid (HA)

Country	Source of HA	Appearance	Declared mol. weight, MDa	Max mol. weight, MDa	Min mol. weight, MDa	Content of low mol. weight fractions, %		
						Up to 10kDa	10–150kDa	Total up to 150kDa
Russia	Animal	powder	0.5	1.2	0.5			
France	Bacterial	flakes	2.6	7.2	2.4	4.8	11.2	16
Japan	Bacterial	granules	2.8	6.4	2.6	5.4	12.6	18
China	Bacterial	granules	1.6	5.2	1.7	9.6	16.4	26
Switzerland	Bacterial	granules	1.2	3.4	0.9	8.0	15.0	23

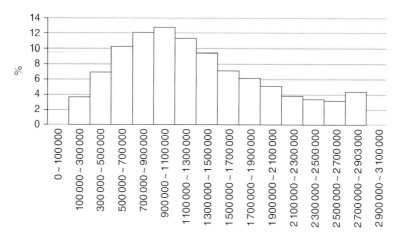

Figure 6.7   *Molecular weight distribution of hyaluronan (according to the Shiseido Co, for the sample HA, lot No. 339)*

Moreover, all studied samples are heterogeneous with respect to fractional composition. The Japanese biotechnological HA production company Shiseido Co. Ltd. reports similar findings of the heterogeneity in the fractional composition of hyaluronan. The molecular weight composition of HA supplied by the Shiseido Co. is shown in Figure 6.7.

The results, presented in Figure 6.7 show the percentage of the high molecular weight fractions of hyaluronan in the commercial HA product. Molecular weight range spans broadly from 100 000 to 3 100 000 Da, while the average molecular weight determined by viscometry is reported to be 2.5 MDa in the certificate of analysis for this product. Manufacturers of medical HA-based preparations should keep in mind that HA supplied by biotechnological companies, if not indicated otherwise, is a mixture of different fractions. Therefore, it is clearly insufficient to indicate the parameter of average molecular weight for such materials. The fundamental characteristic of material for biomedical applications and aesthetical medicine is not an exception here and should include parameter of molecular-weight distribution (polydispersity of the polymer), that is, an information on the fractional composition of polysaccharide. Ideally, depending on the specific application, the finished product should comprise a narrow fraction of HA with a well-defined molecular weight. Such control of molecular weight would make it possible to reduce the number of side reactions after HA-based product administration.

6.1.5   Antioxidant Efficiency of Hyaluronan and other Biologically Active Compounds as Potential Products for Aesthetic Medicine

The following part summarizes the studies dedicated to evaluation of the action of oxidants, including free radicals on skin and the antioxidant efficiency of hyaluronan and other biologically active compounds as potential products for aesthetic medicine [18].

No more than 50 years have passed since the hypothesis of the role of free radicals in ageing of animals and humans evolved. Since then, the hypothesis became the free-radical theory of ageing. The word 'antioxidant' – meaning a compound that slows down the development of destructive oxidation processes – became a permanent word in doctor and cosmetologist vocabularies. The majority of vitamins, amino acids, oligopeptides and mono- and polysaccharides are to some extent able to inhibit the oxidative reactions due to their interactions with oxygen-contained radical species (their other name is 'reactive oxygen species, ROS') as well as influence the rate of the decomposition of structurally different peroxide compounds. It comes as no surprise then, that almost every insert to aesthetic medicine products mentions its antioxidant activity. However, since so many organic compounds have antioxidant properties, it is important to know the quantitative criteria of the evaluation of the efficiency of the antioxidant action. Unfortunately, it is not very common to find a scientific study in which just semi-quantitative evaluation of such action could be presented.

The main mechanisms of the appearance of the active radical forms of oxygen (ROS) in the body are usually related to the distortion of the functioning of the electron transport chains (ETC) of the mitochondria or microsomes. The functions of the mitochondrial ETC are a realization of the subsequent oxidation-reduction reactions of the electron transfer from the substrate of oxidation to the oxygen as a final electron acceptor. At the same time, two-electron reduction of O_2 to H_2O takes place, which is why the free radicals (very reactive species with free valence) should not appear. However, it was shown in [19,20] that the normal electron transfer in the mitochondria (two-electron reduction of O_2) is inevitably spontaneously interrupted, during which only one-electron reduction of O_2 takes place and superoxide ion-radical O_2 appears. This radical is not very reactive, but when we discuss the mechanism of the potential harm of O_2, it is usually referred on the reaction 6.6:

$$O_2^- + H_2O_2 \rightarrow OH^- + OH^* + O_2 \qquad (6.6)$$

This reaction results in an extremely strong oxidative hydroxyl radical OH*, which reacts with practically any organic molecule with an extremely high rate. The radicals O_2 and OH* initiate the reactions of the peroxide oxidation of lipids, and the oxidation of proteins, nucleotides and polysaccharides. They can also cause destruction of DNA helixes and destroy the whole cell [21,22]. In addition to these two radicals, another hydroperoxide radical, HO_2* is related to ROS as well as singlet oxygen. It is common knowledge that for neutralization of the negative effect of ROS, the organism must have the antioxidant defence system. This system contains superoxide dismutases (SOD), catalases, glutathione, glutathione peroxidases, sulfur-containing amino acids, tocoferol and ascorbic acid. The enzyme SOD was found in all aerobic organisms, namely cytozole (Cu, Zn-SOD), mitochondria (Mn-SOD) and in aerobic bacteria (Fe-SOD) [23]. This enzyme catalyses the reaction of dismutation of the anion-radical O_2- :

$$O_2^- + O_2^- + 2H^+ \rightarrow H_2O_2 + O_2 \qquad (6.7)$$

Glutathione peroxidase functions the same way to decompose the hydrogen peroxide into water and oxygen but in addition destroys hydroperoxides by another, non-radical mechanism. Hydrogen peroxide is dangerous because in the presence of the transition metals (e.g. Fe^{2+}) it can be the source of the OH* radical. Vitamin E (alfa-tocopherol) supposedly intercepts the peroxide radicals ROO*, appears at the peroxide oxidation of lipids, transfers them into hydroperoxides ROOH and the vitamin C (ascorbic acid) reactivates alfa-tocopherol. It is at the same time a co-factor of peroxidase (an enzyme ascorbate peroxidase). Glutathione presented in the cells in high concentration, is an OH radical and singlet oxygen acceptor and is a co-factor of glutathione-peroxidase and glutathione reductase [24]. Such a powerful system of the antioxidant organism protection is accumulated mainly in the cell since ROS are produced by mitochondria. Unfortunately, an appearance of the oxygen-contained radicals can happen not only during the natural biological processes, but as a result of the action of unfavourable, aggressive environmental factors. The action of the natural background of ion emitting, UV exposure and photochemical smog, all are sources of increased concentrations of ozone, which leads to an appearance of the same extremely chemically active OH radical in the intercellular substances of the derma and epidermis. This is the reason for premature skin ageing [25]. In this case, an antioxidant system of the organism protection is not effective because the pathologic processes take place on the intercellular level. That is why it is logical the subcutaneous injection of the drugs should provide an efficient antioxidant therapy. In order to better understand what should be included into these products (composition, mutual interaction and quantitative ratio of the different components) it is necessary to know the mechanism of their action.

The basis of the free-radical mechanism of oxidation, which includes combustion and explosion, was created by N. Semenov as part of his fundamental work dedicated to the theory of branched chain reactions (Nobel Prize for Chemistry in 1956). Based on this (now common) theory, the oxidation of hydrocarbons takes place according to the mechanism of the free-radical chain reaction with forced branching. Let's look at the main rules of this mechanism, which can take place in the extracellular skin matrix.

The first stage involves initiation of the oxidation. As was mentioned earlier, as a result of the action of the unfavourable environmental factors, in upper layer of the skin the reactive hydroxyl radical could be formed. In the place of its formation it immediately reacts (the efficiency of the reaction is determined by the rate constant of this process) with the neighbouring biomolecule (RH):

$$RH + OH^* \rightarrow R^* + H_2O \tag{6.8}$$

where RH is any biomolecule of the extracellular skin material.

The bioradical R*, formed as a result of this reaction, can be transferred into peroxide radical ROO* by the reaction with oxygen, which the skin already contains:

$$R^* + O_2 \rightarrow ROO^* \tag{6.9}$$

The probability of such a reaction in skin is quite high, since at the normal conditions O_2 can penetrate into skin to 260–270 µm [26]. The epidermis, papillary layer and derma upper layer are supplied with air oxygen. The ability of the skin to consume oxygen does not reduce with age.

The following 'destiny' of the peroxide radical depends on its chemical structure, but as a rule, this radical particle is known to be involved in concur reactions of disproportionation, isomerization or atomic H removal from the neighbouring biomolecule:

$$ROO^* + RH \rightarrow ROOH + R^* \tag{6.10}$$

For example, polysaccharides are known for creating peroxide radicals of two types [27]. The first is created as a result of the attack of O_2 of free valence on the atom C1, which was created at the breaking of the glycosidic bond. The less stable peroxide radical of the second type is created as a result of the reaction of oxygen with a non-paired electron and localized on any middle carbon atom of the pyranose cycle.

The end of these processes leads to oxidative destruction of the polysaccharide. The main chain is broken, which in turn results in the formation of low molecular weight oxygen-containing products. Alongside the destruction processes, the linkage reactions are characteristic for protein molecules like collagen, these reactions lead to formation of intermolecular peroxide bonds. All of these processes inevitably lead to degradation of the extracellular matrix.

The reaction (6.10) is the initial step of the development of the chain free-radical oxidation process. The radical R*, which is created in the reaction, once again goes to the reaction (6.9), but hydroperoxide ROOH can decompose according to the radical mechanism depending on the reaction conditions:

$$ROOH \rightarrow RO^* + OH^* \tag{6.11}$$

This stage is a stage of the forced chain branching (i.e. it leads to appearance on the new OH* radical), which is an initiator of the chain process of carbohydrate oxidation. The macro-radical RO* is usually less active and could cause the chain to break according to the reaction (6.12) with formation of the peroxide linkage:

$$RO^* + RO^* \rightarrow ROOR \tag{6.12}$$

Based on this already classical free-radical chain oxidation mechanism, it is obvious that it is possible to inhibit this process by introduction of the substances that, from one hand concur with RH for the binding with hydroxyl radical in the reaction (6.8), and on other hand concur with RH in the reaction (6.10), which does not form the macroradical R* or lead to transformation of the hydroperoxide ROOH without formation of the radical OH*. It means that the reaction (6.11) cannot go through the radical mechanism. For an efficient antioxidant the following equations must be followed:

$$K(AH + OH) \times [AH] \gg K(RH + OH) \times [RH] \tag{6.13}$$
$$K(AH + ROO) \times [AH] \gg K(ROO + RH) \times [RH] \tag{6.14}$$

where $K(AH+OH)$ – Rate constant of the reaction of the antioxidant (AH) with OH*, $M^{-1}s^{-1}$;

[AH] – antioxidant concentration, M;
$K(RH+OH)$ – rate constant of the reaction (6.8), $M^{-1}s^{-1}$;
[RH] – biomolecule concentration, M;
$K(AH+ROO)$ – rate constant of the reaction of antioxidant with peroxide radical, $M^{-1}s^{-1}$:
$K(ROO+RH)$ – rate constant of the reaction 6.10, $M^{-1}s^{-1}$.

If all parameters in the equations (6.13) and (6.14) are known, it is possible to quantitatively evaluate antioxidative efficiency as a function of concentration of the introduced antioxidant [AH]. The *in-vivo* antioxidative evaluation is extremely difficult and is almost always an impossible problem. In order to perform the model experiment the authors in [18] applied the methodology, which was successfully used for the study of the radiation-chemical oxidation of different organic compounds [28–33]. That is why the application of the methods of radiochemistry was not unintentional. The US scientist Hartman, the author of the hypothesis on the role of free radicals in human and animal ageing formulated in 1954, gave the title of his first article dedicated to this subject 'Aging: theory, based on free-radical and radio-chemistry'.

This chapter will describe the products of the biologically active compounds related to the different classes: glycosaminoglycanes – hyaluronan (as a sodium salt); sulfur-containing amino acids – L-cysteine, L-methionine and oligopeptides – L-glutathione; vitamins – ascorbyl-phosphate (derivative of vitamin C). The exposure was done by gamma irradiation of Co^{60}. The detailed experimental methodology was similar to that mentioned in [28–33]. Two series of the experiments were performed. In the first series, the role of the additives of investigated compounds on the radiochemical oxidation of polyamide was studied. (The aliphatic polyamide is considered in many experiments in radiochemistry as an analogue of a protein molecule.)

Figure 6.8 shows the kinetic curves of the accumulation of the hydroperoxides ROOH, which form as a result of the oxidation of polyamide under the action of ionized exposure in the conditions of pure sample of polymer and with different bio-additives. At the polymers exposure in the presence of oxygen, the same processes (6.9–6.12), which were considered earlier, take place. The difference is that the initiating stage of reaction (6.8) has been

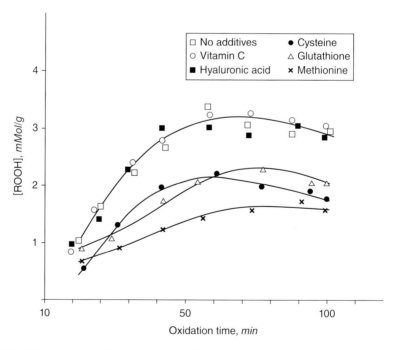

Figure 6.8 *Kinetic curves of accumulation of hydroperoxides in the exposed polyamide with different additives*

replaced by the appearance of macroradical R* at the direct action of the exposure. The results, obtained at the experiments, show that hyaluronan and vitamin C have practically no influence on the rate of hydroperoxide accumulation in the exposed polyamide. At the same time, glutathione and even more so, methionine and cysteine, significantly reduce the amount of the generated hydroperoxide. This likely happened due to the presence of sulfur contained fragment, which leads to partial decomposition of ROOH through a non-radical mechanism. In the result of such decomposition, hydroperoxide produces the hydroxyl ion OH- instead of the active hydroxyl radical OH*. The hydroxyl ion OH- cannot induce the subsequent development of destructive oxidation processes. Thus, the series of the experiments showed that sulfur containing amino acid and tripeptide glutathione in the experimental conditions effectively slow down polymer oxidation at the stage of hydroperoxide destruction. At the same time, HA and sodium ascorbylphosphate do not influence on the rate of the radiation oxidation of aliphatic polyamide, that is, do not reveal any antioxidant activity.

The second series of the experiment was directed toward the study of the efficiency of the interaction of the studied compounds with OH radical. In order to do it, the method of radical competing acceptors is been used. The basis of this method is study of radiochemical destruction of the dissolved compound (indicator) in presence of the studied compound. Using this method, the reactivity of a large amount of compounds of biological origin when exposed to radicals have been studied, particularly biopolymers [27].

In this series of experiments, the neutral aqueous solutions of the organic dye Rhodamine 6 g was prepared in the concentration 2.5×10^{-4} M. These solutions were saturated with nitrous oxide N_2O and treated with exposure to gamma irradiation. Under the action of the ionized exposure on the saturated aqueous solution of N_2O, practically the only radical particle formed under exposure is the OH radical [46]. If the solution contains organic dye, in our case Rhodamine 6 g, then the result of the reaction with OH* will be its decomposition, which will be shown as the solution decolouration. The level of decolouration could be measured spectrophotometrically at the dye absorption maximum of 545 nm. The experimental results were used for kinetic curves of the radiation decolouration at the different exposure time. The initial linear fragments of these curves determined the rate of the dye radiochemical oxidation. In radiochemistry such a kinetic parameter is marked as a radiochemical yield of the dye decolouration G, which is equal to the number of decolourized molecules on 100 eV of the absorbed energy of ionized exposure. At the addition of different biologically active compounds to the aqueous solution of dye, the effect of the rate of radiochemical oxidation of Rhodamine 6 g was evaluated. The molar concentration of the added compounds was similar and equal to concentration of dissolved Rhodamine 6 g. Table 6.2 shows the quantitative results of the conducted study.

The results presented in the table show that all studied compounds quite efficiently protect the dye from decolourizing, a trait related to the introduced additives. The most antioxidation action is shown for glutathione, cysteine and vitamin C; each reduces the rate of the oxidative dye degradation two-fold. Introduction of glutathione, cysteine and vitamin C all together leads to the synergetic effect and slows the process of the decolorizing by almost three times. This common effect of these components could be explained by the fact that glutathione and cysteine can reduce the oxidized form of the ascorbic acid.

Table 6.2   *The radiochemical yield of the dye decolouration in presence of different bioadditives*

#	Aqueous solution of Rhodamine, 6 g	Radiochemical yield of dye decolourizing, molecules/100 EV
1	No additives	2.06
2	Glutathione	1.12
3	Hyaluronic acid	1.45
4	Cysteine	1.05
5	Methionine	1.20
6	Vitamin C	1.02
7	Glutathione + Vitamin C	0.81
8	Glutathione + Cysteine + Vitamin C	0.70

At the same time they transform into disulfate compounds, which also have large affinity to hydroxyl radical.

From all studies of biologically active compounds, hyaluronan reveals the lowest level of antioxidative properties. This correlates with the fact that rate constant of the reaction of OH* with sodium hyaluronate is approximately three times lower than values of the rate constants of glutathione, sodium ascorbate and cysteine [18]. Based on the data mentioned previously, we try to evaluate the maximum antioxidative effect from the introduction of the studied compounds into derma.

Imagine that 3 ml of the aqueous solution of the product with the concentration of the active compounds of 1 weight % (10 mg/ml) injected all at once on the injection zone of $100 \, cm^2$. If the dermal layer has a width of 2–4 mm, we can find that the molar concentration of antioxidant (on condition of equal distribution of the product in the injection zone for some time) is approximately $10^{-3} M$.

The rate constant K(AH+OH) for glutathione, cysteine and ascorbate is about 1010 $M^{-1} s^{-1}$. The concentration of endogenous hyaluronic acid in derma is approximately $10^{-3} M$ [33] and collagen is almost 10 times higher. The rate constant of the reaction (6.8) for hyaluronan is 5×10^9 and for collagen proteins 1×10^8 [7]. If the previously mentioned data will be put into equation (6.8), the simple kinetic scheme will allow one to properly evaluate the upper limit of the antioxidation effect. In the case of hyaluronan the rate of the oxidative degradation can be reduced in three times, and for collagen, in almost 10 times. Obviously, such estimation is based on the simple model and is quite approximate. However, even if the *in-vivo* effect after addition of antioxidants is lower, the value of the antioxidation therapy is nevertheless hard to under evaluate. Such protection from oxidation is extremely important for collagen since it is related to so-called 'low exchange proteins'; its half-life is weeks or even months. For comparison, the half-life of glycosaminoglycans is significantly lower (for hyaluronic acid in skin, only 1–2 days). For such long periods of time, collagen is able to collect a significant amount of defective structures, which will affect the quality of the dermal extracellular matrix.

Finally we would like to note, that the complex injection of glutathione, cysteine and ascorbic acid derivatives is quite an efficient means of antioxidation whose mechanism of action is revealed at the different stages of the free-radical chain process of the biomolecules oxidation.

6.1.6   Bio-Repairants as a New Class of Injectable Products Based on Hyaluronic Acid Modified with Low Molecular Weight Bio-Regulators

Skin ageing as a particular aspect of an entire organism ageing can be viewed as a change in the steady state when complex compounds like collagen, elastin and glycosaminoglycans including HA are continuously synthesized in the course of one processes and decomposed in the course of others. The coordination of synthesis and breakdown processes is extremely important for maintaining skin health. Thus, for instance, there is a clear correspondence between the intensity of the metabolism of glycosaminoglycans and dermal collagen proteins. The intensity of skin ageing largely depends on the prevalence of one of these processes. With ageing, or under the action of unfavourable environmental factors, the changes in intensity of these processes lead to a decreasing rate of formation with an increasing rate of biomolecule breakdown. From this standpoint, the strategy realized in injection cosmetology should be directed toward the creation of physiologically favourable conditions for strengthening the metabolic activity of skin cells, which would lead to the activation of the synthesis of the basic components of the extracellular dermal matrix. In order to realize such a strategy, Russian scientists developed the technology of solid-phase modification of polysaccharides, particularly hyaluronan, by different low-molecular weight bio-regulators (amino acids, vitamins and oligopeptides) [34–39]. The functional –OH, –COOH, –NH$_2$ and –SH groups in the structure of low-molecular weight bio-regulators make it possible to 'graft' required biologically active compounds onto HA macromolecules under conditions of mechano-stimulated reactions, that is, to perform chemical immobilization. It is important to note here that the unique property of HA, which makes it possible to realize the targeted delivery of active ingredients, is the bio-recognizable motive in the structure of a polysaccharide macromolecule that can interact with the cellular surface of fibroblasts (as already was mentioned, the polysaccharide binds specific protein receptors CD44 and RHAMM located on the surface of cytoplasmic membrane).

The method of solid-phase modification is based on intensive mechanical processing of the solid mixture of reagents, conducted in the absence of solvents and diluents under the joint action of pressure and shear deformation using the equipment of different types (see Section 5.2.7 in Chapter 5 for more detail). It is paramount to note that the technology permits chemical reactions to be carried out with different low-molecular weight bio-regulators without the use of bi-functional reactants. Bioactive compositions comprising HA grafted (immobilized) with vitamins, amino acids and oligopeptides are obtained in a single-stage technological regime. The active ingredients are then covalently bound to the biopolymer macromolecule. Figure 6.9 summarizes the compounds of different classes that are 'grafted' onto the hyaluronan macromolecule [25,40].

By its very nature, the process of chemical bond formation between hyaluronan macromolecules and organic compounds under the joint action of pressure and plastic deformation is similar to the reaction of polypeptide synthesis from amino acids or polysaccharides formed from simple sugar. The formation of chemical bonds can be visualized as a joining of –OH, –COOH and –NH$_2$ groups accompanied by removal of the water molecule. However, in aqueous mediums, equilibrium of this type of reaction is shifted towards parent substances rather than the products of the reaction. Therefore, such reactions, both in living systems and under laboratory conditions, are achieved as a result of a complex multistage process, far from being similar to a simple process of water removal. The proposed solid-phase

Structural Formula	Name
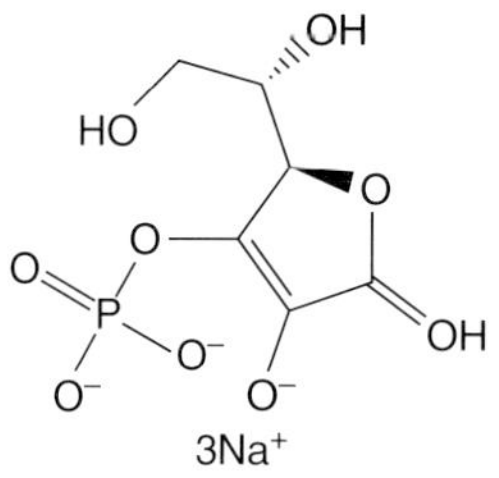	Sodium (or magnesium) Ascorbyl phosphate (Vitamin C derivative)
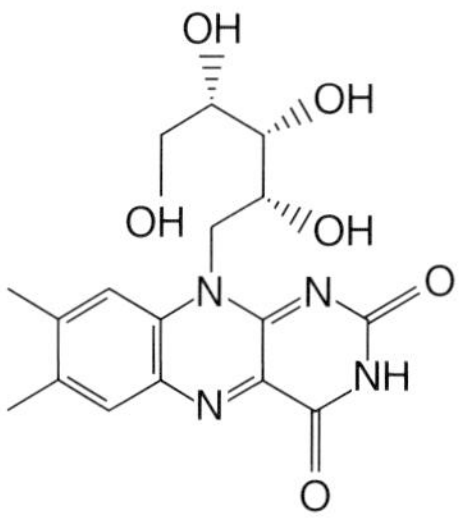	Riboflavin (Vitamin B$_2$)
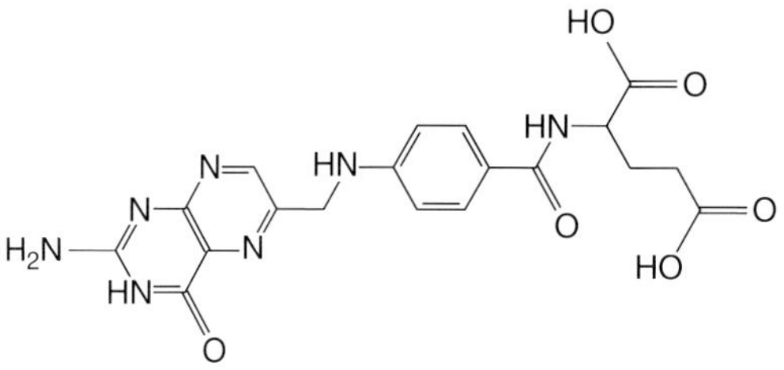	Folic acid (Vitamin B$_9$)
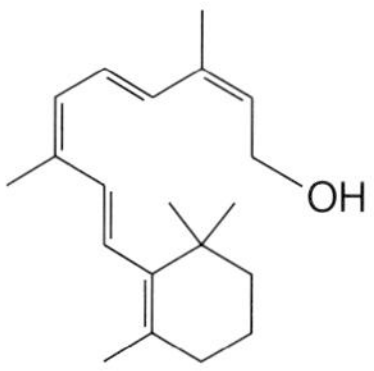	Retinol (Vitamin A)
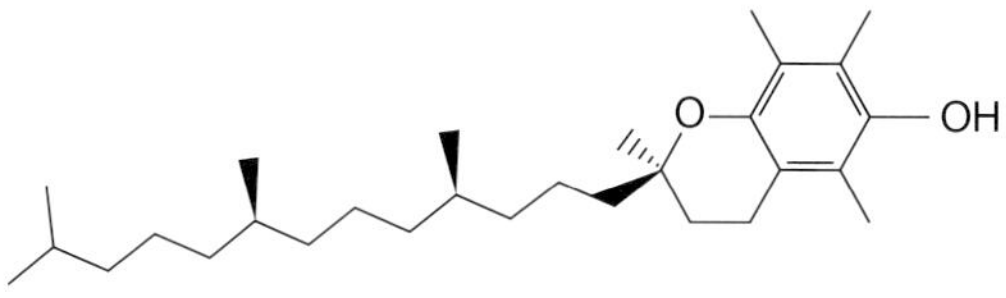	α − Tocopherol (Vitamin E)

Figure 6.9   *Low molecular weight bio-regulators used for solid-state modification of hyaluronic acid*

Structural formula	Name
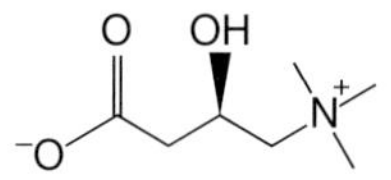	L−carnitine (Vitamin BT)
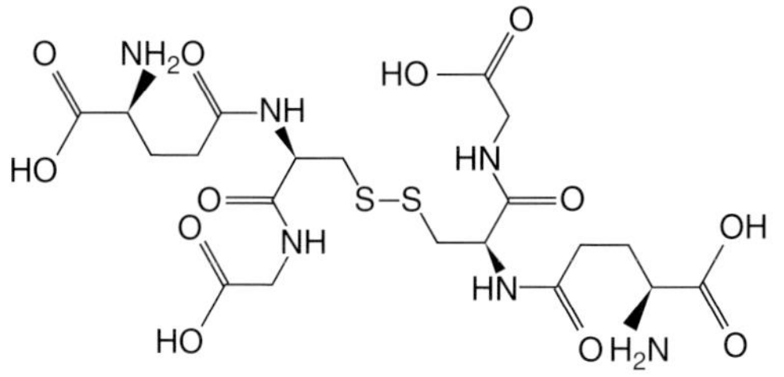	Glutathione (tripeptide−glutamyl− −cysteinyl−glycine)
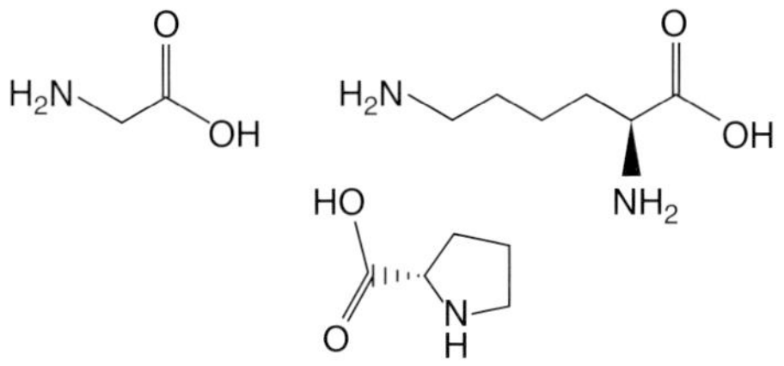	Amino acids−Glycin, Proline, Lysine
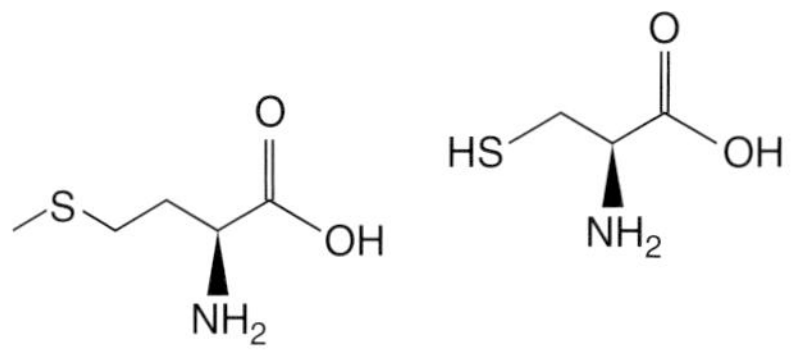	Sulphur containing amino acids− Methionine, Cysteine

Figure 6.9 *(Continued)*

procedure makes it possible to carry out the necessary process in one stage, without the use of bi-functional toxic components. The solid-phase chemical immobilization can be compared to the sewing of beads to a cloth: the molecule of the biologically active ingredient is attached to HA through a stable covalent chemical bond and the attached molecule seemingly dangles on a small 'filament', forming a large complex that in essence represents a unique macromolecular 'depot' of biologically active material in the place of injection. The mobility of the polysaccharide chain in the modified HA is limited compared to the native structure. It is no longer easy to 'disentangle' and to depolymerise the complex repeating units by active hyaluronidases. This leads to an increased residence time of formulation in the dermal layers. Further, small filaments are hydrolysed and break down, releasing the bioactive component that may remain in the zone of injection for a sufficiently long time and supply necessary vitamins, amino acids and oligopeptides in constant concentrations.

Using the technology of solid-phase modification, the product line called 'Hyalrepair' was created [41,42] to include 10 different bioactive compositions of densely cross-linked Na^+, Cu^{2+} and Zn^{2+} salts of HA with chemically immobilized vitamins (ascorbic and folic acids and riboflavin), amino acids (glycine, proline, lysine, valine, carnitine, cysteine and methionine) and oligopeptides (glutathione). It should be mentioned that hyaluronan, in these formulations, plays the role of 'actively' transporting important biological compounds to the targeted cells of the organism. One of such compounds is ascorbic acid (Asc) whose value for the health of skin cannot be overestimated. Asc has a basic biological function that determines the integrity of connective tissue: with the formation of hydroxyproline and hydroxylysine, it participates in the metabolism of glycine, tyrosine and hydroxylation of proline and lysine residues in proteocollagen. Proline hydroxylation is necessary for stabilization of the triple helix of collagen, while the hydroxylation of lysine is very important for subsequent covalent bonding between collagen molecules that takes place with the formation of collagen fibrils. This fibril producing process determines stabilization of the extracellular connective tissue and reduction of capillary permeability. Collagen synthesized under conditions of vitamin C deficiency proves to be significantly deprived of hydroxyl groups and O-glycosyl residues. These structural alterations prevent the formation of strong and functional fibres and are the reason for frequently seen skin defects. Furthermore, as an inhibitor of melanogenesis, vitamin C blocks the action of tyrosinase, the key enzyme of the melanin formation and in so doing blocks the synthesis of melanin, reducing dopa-chromium into dopaquinone.

According to the new functional classification, vitamin C belongs to the group of 'potent' vitamin-antioxidants. The antioxidant action of vitamin C consists of two processes: (1) the effective binding with hydroxyl radicals, which leads to the inhibition of radical chain oxidation process; and (2) its collaboration with α-tocopherol (vitamin E), where vitamin C reduces the oxidized form of vitamin E. Similarly, vitamin C converts folic acid (vitamin B9) into a bioactive state. There is information in scientific literature about the ability of vitamin C to influence the formation of glycosaminoglycans, in particular HA and chondroitin sulfate, and to stimulate the proliferation of fibroblasts [43]. Physiological action of ascorbic acid is not only due to its ability to stimulate collagen synthesis but also because of inhibition of the production of metalloproteinases, the enzymes that destroy dermal collagen [44]. Numerous published works have confirmed ascorbic acid has the ability to improve healthy skin condition, including effective reduction of primary signs of ageing. Vitamin C belongs to the class of water-soluble vitamins and therefore is rapidly eliminated from the organism without any appreciable accumulation. In the case of the 'Hyalrepair' products, the solid-state technology makes it possible to chemically 'graft' up to 95% of ascorbic acid weight onto a HA macromolecule, thus creating an active 'depot' of vitamin in the site of injection that lasts a sufficiently long time [34].

The amino acids glycine, proline, lysine and valine are components of the main proteins of the extracellular dermal matrix. The inclusion of these amino acids in the composition of 'Hyalrepair,' along with other low-molecular bio-regulators and microelements, is essential for triggering synthesis of dermal collagen and elastin, which are extremely important for achieving steady, long-term effect.

Sulfur-containing amino acids like cysteine, methionine and tripeptide glutathione are very powerful antioxidants that participate in different stages of free-radical chain reactions of biomolecule oxidation [18]. For example, cysteine participates in the synthesis of taurine, the substance that effectively blocks the peroxide oxidation of lipids by binding hypochlorite anion to form chloramine complex. In any organism, cysteine and glutathione reduces the oxidized form of vitamin C to its initial active form while methionine (being an essential amino acid) is metabolically related to cysteine. To strengthen the antioxidant properties, certain preparations of 'Hyalrepair' are enriched with folic acid, the OH-radicals acceptor, which is especially effective in the presence of vitamin C. Other biochemical functions of folic acid as an enzyme component include the ability to transfer the single-carbon radicals, that is, formyl, hydroxymethyl, methyl, methylene, methine and formimine, and participate in the synthesis of the amino acids like serine or methionine. The biological role of riboflavin (vitamin B) consists in the stabilization of the extracellular matrix of connective tissue. The molecule also facilitates absorption of oxygen by skin cells and accelerates the *in-vivo* transformation of pyridoxine into its active form.

Besides its main role as a carrier of active forms of fatty acids through the membranes during lipotrophic oxidation, the presence of carnitine contributes to the normalization of the water–salt balance in the skin. Microelements in the form of Cu^{2+}, Zn^{2+} and Mg^{2+} cations in the complex with proteoglycans and glycosaminoglycans of the extracellular matrix determine the swollen state of the skin. Copper is a component of the extracellular copper-containing enzyme lysyl oxidase that participates in the formation of intra- and inter-chain cross-links in collagen and the elastin. Cross-link formation is disrupted when availability of copper is scarce, leading to reduced strength and elasticity of collagen fibres. Copper is often more effective when paired with zinc. In the human organism, zinc is found primarily in the skin where this microelement acts as a component of 70 enzymes, the majority of which participate in the processes impeding degradation of the extracellular dermal compounds.

The authors who developed the technology of the solid-phase modification of hyaluronan by low-molecular bio-regulators consider the 'Hyalrepair' line of products to be non-medicated macromolecular therapeutic agents (NMMTA), while their administration is carried out within the framework of the complex 'Bio-repair' program. The 'Bio-repair' program implies injection administration of NMMTA with the aim of activating cellular metabolism in the skin for the purpose of restoring damages of the extracellular matrix followed by preventive protection of dermal components from unfavourable and aggressive environmental factors. Injections of NMMTA shift the physiological balance of the metabolic processes established between the cell ensembles and the extracellular dermal matrix and modulate the steady state in this system. The invasiveness of the NMMTA administration, especially in mesotherapy, can develop a reaction similar to that of skin injuries but, obviously, with less intensity and on a smaller scale. Damage of different cell types in the epidermis and derma triggers uncontrolled secretion of lysosomes and lysosomal enzymes into the extracellular matrix. This starts an entire cascade of intercellular interactions consisting of a series of coordinated reactions in different cell types in the damaged tissues and orchestrated by local mediators that control growth factors. This leads to the intensification of biopolymer breakdown processes, that is, cleavage of HA by hyaluronidase, splitting of sulfated glycosaminoglycans from proteoglycans, proteolysis of proteins and so on but also promotes processes opposite to destruction and directed toward the restoration structures of extracellular matrix and reparative regeneration. As a consequence, according to the law of excessive compensation, reparative processes in the cells not only restore damage but also lead to the renovation ('rejuvenation') of cell structures and extracellular substances and allow the balanced synthesis-disintegration processes to proceed at a higher rate. HA modified by low-molecular bio-regulators is a source of the biologically active compounds necessary for completion of bio-repair processes. At the same time the processes are developed against the breakdown of biopolymer structures by free radicals.

To determine the duration of resorption and tissue reaction after intradermal administration of 'Hyalrepair' gels into the interscapular region, animal testing was carried out in white rats in the laboratory of experimental pathomorphology. The periods of observation were from 1 to 30 days. Two very important conclusions were made following the study [41]. First, tissue sections from the site of injection did not contain large conglomerations of gel, but is instead comprised of small cavities, each of which is filled with the gel and encased by a poorly developed capsule. On the micrographs, the gel material is seen in the form of small fragments of a basophilic substance (Figure 6.10 and 6.11).

In some animals, neither capsules, nor microcapsules surrounding the gel fragments were detected. Thus, the introduced preparation is considered bio-inert, as it is not encapsulated in the tissue but rather spread within it and incorporated in the extracellular matrix (Figure 6.12).

Modified HA is a unique 'nano-container' of biologically active compounds, both at the level of the individual macromolecule and small evenly cross-linked cellular structures. This is a kind of reticular structure composed of nano fragments of 50–300 nm. These structures could be studied by the sol-gel method based on the statistical theory of cell-like polymer formation.

The second conclusion from the study is that administration of all tested preparations led to a noticeable proliferation of fibroblasts in the form of large active cells with relatively abundant cytoplasm. After injection of the 'Hyalrepair-02' composition, which includes the amino acid complex, formation of collagen fibres is observed.

The clinical trials of the 'Hyalrepair' line of products were carried out in the Department of Derma-Oncology and Laser Surgery of Central Clinical Hospital of Russian Academy of Science and the Institute of Plastic Surgery and Cosmetology. Measurements of skin

Figure 6.10 *Micrographs of tissue section 3 days after 'Hyalrepair-02' injection. Non-uniform distribution of thick filament structures in the gel is observed. Newly developed capillaries can be seen in the upper part of the image. Magnification 400×*

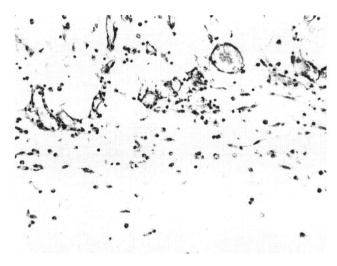

Figure 6.11   *Micrographs of tissue section 3 days after 'Hyalrepair-10' injection. The boundary zone between fat tissue and gel is shown. An increased number of fibroblasts and fully developed blood capillaries are detected. Magnification 400×*

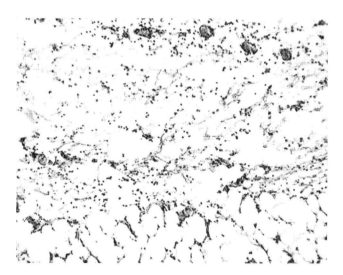

Figure 6.12   *Micrographs of tissue section 7 days after 'Hyalrepair-02' injection. The upper part of the image shows the gel with ingrown fibroblast and blood vessels. The lower part shows the fat tissue. Connective capsule surrounding the gel implant is not detected*

elasticity, pH and hydration levels were carried out to confirm the effectiveness of the preparations. The measurements of skin elasticity and hydration in patients were conducted prior to the administration and after 2, 3 and 4 weeks post-introduction.

In 90% of patients, skin elasticity was improved by 10–12% on average compared to initial value and remained at this level for 3 months after procedure completion. The ageing skin is known to feature reduced viscoelastic properties as a result of qualitative and quantitative changes in dermal collagen proteins. The injection of 'Hyalrepair' preparations makes it possible to considerably improve viscoelastic characteristics of the skin, which indirectly indicates that stimulation of the collagen and elastin syntheses has occurred. Two weeks after the first injection, the skin hydration level in all subjects increased by 6–8%, while after 4 weeks the increase was 12–15% compared to the initial level at the constant pH. This indicates that HA that has been modified in the solid state is retained in the site of injection for at least one month, acting as a source of essential vitamins, microelements and amino acids included in the preparation. In conclusion, it is important to note that the innovative technology of the solid-state modification of the biopolymers allows the novel products with modified HA by low molecular weight biomodulators to be obtained in the single-step technological regime without usage of foreign technological additives.

When the process is conducted in the solid state in the absence of the solvents, insoluble or partially soluble bioactive components such as riboflavin or folic acid can be attached to a macromolecule HA, thus transferring them into a water-soluble form suitable for injection. Biologically active components with the macromolecule HA in the chemically bounded state acquire storage stability, which is extremely important for unstable compounds such as vitamin C and amino acid cysteine, which could be easily oxidized under the usual storage and sterilization conditions. With the process of 'grafting' of the bio-regulators, molecular mass distribution of HA simultaneously became more homogeneous, which reduces risk of unwanted side effects related to non-homogeneity of the fractional composition of hyaluronan.

The innovative product 'Hyalrepair' could also be used for prophylaxis and non-injection therapy of the metabolic diseases and syndromes as topical products. Composition variations of the Hyalrepair line consist mainly of the natural metabolites, which need to be delivered to targeted tissues and organs. Transcutaneous permeability is usually low and is acceptable for the treatment by low and ultra-low doses. Today the pharmacology of ultra-low doses is recognized as a promising trend in medicine. In includes hormones and regulatory peptides, which are a part of a complex system of the specific signalling molecules (cytomedines) – information mediators between cells. The most important characteristic of cytomedines is their high affinity to receptors.

For example, the dissociation constant (K_d) of the hormone complexes with receptors have values in the range 10^{-8}–10^{-10}M [31]. This is why the extra low concentrations of the signalling molecules have a significant physiological or biochemical effect. Hyaluronan often shows the features of cytomedines. It has specific receptors and thus is a direct participant of the information-exchange processes on the cellular and intracellular levels.

Fast decomposition is another important criterion of the signalling molecules that could be attributed to hyaluronan. The fragments of the macromolecule HA have a wide range of structural and conformational states and maintain the ability to bind with receptor CD44 if the fragment contains at least 20 monomers.

It is possible to suggest that the hyaluronan fragments with different molecular weight and structural conformations have different abilities to transfer the intercellular and extracellular signals compared with high molecular weight HA. The polypeptide signalling molecules are protected from the fast degradation by the special protein carriers. The classic example is complexes of oxytocin and vasopressin with neurophysins [45]. It is interesting to note that the cyclic octapeptide vasopressin (the neurohypophysis hormone) together with hyaluronan creates a biological system with the ability to maintain water content in the body.

In this regard, the pituitary hormone vasopressin helps water retention in the body by its reabsorption in the renal tubules and thus controls expression of the gene hyaluronan synthase-2 in the kidney [46].

'Grafting' different bioactive components onto hyaluronan improves the stability of the complex. With assistance of the CD44 receptors, such complexes have ability of prolonged transportation into cells missing key metabolites using the endocytosis mechanism.

Today, medicine uses ultralow doses of antibodies for the treatment, prevention and rehabilitation of various diseases. The innovative technology allows for the synthesis of the bio-repair product line ('Hyalrepair'). These products may find wide application in medicine, particularly for the prevention and therapy of metabolic abnormalities that require correction of metabolism by low and ultra-low doses of the natural metabolites.

It is known that different damaging factors can result in the same symptoms. Typically, the symptoms are related to the integrative processes in the cell and result as an imbalance of the three main types of metabolism: carbohydrate, protein and lipid [47]. It is often possible to observe effects such as bifurcation, futile cycles and 'vicious circles' in the metabolic pathways, all of which be eliminated by the delivery of physiological doses of natural regulators and/or metabolites. Thus, systemic approaches to any pathology places importance in finding the links between the different types of metabolisms as well as their integrative relationships with low and ultra-low doses of key metabolites. Hyaluronan-based products modified with bio-regulators of low molecular weight can serve as an ideal means to achieve these goals.

An analysis of the literature on administration of low-and ultra-low doses of biologically active compound reveals two distinct areas of study. The first examines signalling molecules and their receptors and the second explores low and ultra-low doses of key metabolites. The scientific basis of physiological effects of ultra-low doses of metabolites was established by E.B. Burlakova et al. in 1986. They published an article that describes the unusual dose-effect dependence and showed that equal physiological effects can be reached both in the field of ultra-low concentrations of antioxidants, and at higher near-physiological concentrations. The intermediate concentration did not cause any effect. It is quite remarkable that the difference between the effective concentrations is in the several orders of magnitude [48]. It has been found that such a bimodal effect can be revealed by the substances with a particular electronic structure of the molecule [49].

In order to perform the detailed studies on such effect, a series of 35 chemical compounds (picolinic acid derivatives) were synthesized. All compounds were studied in a wide range of concentrations, from ultra-low to physiological level. From this series two compounds were identified. These compounds showed a significant bimodal activity that affects the rate of plant cell proliferation [50]. In order to find the correlation between the electronic structure and the physiological properties of the two compounds with a distinct bimodal activity, quantum-chemical methods were used [51]. These studies resulted in the hypothesis that the mechanism of the bimodal effect is associated with the electron-acceptor effect of pyridine cycles. An attachment of the electron-donor substituents to the pyridine leads to an energy increase of the highest occupied molecular orbitals of the pyridine ring, and consequently to decrease the ionization potential. For example, it was shown that the amino groups could be such substituent [51]. In the reaction chain of biochemical synthesis of hyaluronan, the key (rate-limiting reactions) stage is the synthesis of glucosamine from fructose-6-phosphate. The source of amino groups in this reaction is glutamine. The transfer to the amino group from glutamine to fructose-6-phosphate competes for glutamine with several other reactions in which glutamine is used for the synthesis of purine and pyrimidine nucleotide bases for synthesis of DNA and RNA, as well as certain amino acids, and so on. Therefore, glutamine is a key metabolite in 'crossroads' of the synthesis of nucleic acids, proteins, hyaluronan and other glycosaminoglycans. A reduced level of glutamine in the blood and tissue leads to slower syntheses of the biopolymers and imbalance of the processes of synthesis decomposition of biopolymers. Such an imbalance in the synthesis and decomposition of HA primarily occurs in tissues where there is no circulatory system, namely the skin epidermis and cartilage tissues. This is one reason for different pathologies. These data reveal a very interesting perspective on the 'design' of the products based on solid-state modified hyaluronan. Such products could be active in very low doses and/or have a bimodal activity, which is able to 'extinguish' the imbalances in metabolic pathways and remove the biological systems from the 'vicious circle' in various organs and tissues.

6.2   Hyaluronan in Arthrology

As was noted before, the reducing ability of cartilage tissue chondrocytes to synthesize glucosamine is one of the main factors of osteoarthritis development. The healthy knee contains about 2 ml of synovial liquid and the hyaluronan concentration in it ranges from 2.5–4 mg/ml (0.25–0.4%). Patients with osteoarthritis have a 2–3 times lower rate of synthesis and concentration of hyaluronan in synovial liquid compared to the normal rate, reduction of the molecular weight of HA due to accelerated hydrolysis and reduction of viscoelastic properties [52]. As a result, the friction of the surfaces of the cartilage increases as well as destruction of cartilage and bone tissue [53–55]. That is why world medicinal practice considers preparations of glucosamine with chondroitine sulfate (see Table 6.3) to be a first rate product for regeneration of cartilage tissue of the joints through activation of HA synthesis by delivery of the precursors subcutaneously or orally. The other approach of the biomedical technology of the osteoarthritis treatment is direct injection of the products of high molecular weight or cross-linked hyaluronic acid (see Table 6.4). There is a significant amount of scientific, patent and medicinal literature with positive results from the treatment of osteoarthritis by these two technologies [56–60]. There is also plenty of evidence that confirms efficiency of the intra-joint injection of hyaluronan in reduction of the pain syndrome and improving of the functional activity of the joints at gonarthrosis and coxarthrosis. It was shown that application of HA helps to recover the properties of the synovial liquid, suppress synthesis anti-inflammatory cytokines and prostaglandins, activate anabolic processes and reduce katabolic processes in the cartilage tissue [61–67]. Table 6.3 shows several hyaluronan based products with structure-modification activity. These products are used for treatment of osteoarthrosis. Most of them are formulated as a powder or tablets and are intended for oral delivery.

Other products, presented in Table 6.4, are used for treatment of osteoarthritis. All of them are formulated as solutions in prefilled syringes for injections.

Glucosamine (Table 6.3) and hyaluronan products (Table 6.4) belong to the category of structure-modifying compounds that, contrary to non-steroidal anti-inflammatory drugs, not only stop the pain syndrome but also help to recover cartilage tissue.

Table 6.3   *Hyaluronan products for treatment of osteoarthrosis*

Product	Active Ingredient	Manufacturer	Drug Formula
DONA	Glucosamine sulfate	Rottafarm (Italy)	2 ml solution (200 mg/ml) for intramuscular injection
DONA	Glucosamine sulfate	Rottafarm (Ireland)	1.5 g powder for preparation of oral solution
Aminoarthrin	Glucosamine	Moscow Pharmaceutical Company (Russia)	300 mg tablets
Structum	Chondroitin sulfate	Pierre Fabre Pharmaceuticals (France)	500 mg and 250 mg capsules
Chondroitin-AKOC	Chondroitin sulfate	Sintez (Russia)	250 mg capsules
Artra Chondroitin 750	Chondroitin sulfate	Unipharm Inc. (USA)	750 mg coated tablets
Artra	Chondroitin sulfate + glucosamine hydrochloride	Unipharm Inc. (USA)	Coated tablets
Theraflex	Chondroitin sulfate + glucosamine	Sagmel Inc. (USA)	Capsules
Alflutop	Natural bioactive concentrate	BIOTEHNOS S.A. (Romania)	10 mg/1 ml ampule for injection
Piascledine	Herbal extract	Laboratoires Expanscience	300 mg capsule

Table 6.4   *Hyaluronan products for treatment of osteoarthritis*

Product	Active ingredient	Manufacturer	Drug formula
Viscorneal ortho	High molecular weight, high purity sodium hyaloronate 1% solution, pH 7.0–7.5	CORNEAL (France)	2 ml syringe
SYNVISC® (hylan G-F 20)	Sterile viscoelastic gel containing Hylan A (sodium hyaluronate of 6 MDa) and hylan B (hydrated gel (pH 7.2 ± 0.3)	Genzyme Corporation (USA)	2 ml syringe
Syinocrom	Highly purified hyaluronic acid 1.6 MDa	Croma (Austria)	2 ml syringe
Suplasyn	Sodium hyaluronate – 20 mg, phosphate buffer – 2 ml	Bioniche Pharma Group Limited, (Ireland)	2 ml syringe
Ostenyl	1% gel of viscous gel of hyaluronic acid with molecular weight 1.2 Da	TRB Chemedica Ltd (UK)	2 ml syringe
Viscoseal	Sodium hyaluronate 0.5% solution	TRB Chemedica Ltd (UK)	10 ml syringe

The mechanisms of the HA action on the molecular level have been intensively studied. It has been established that when high molecular weight HA is added into the cell culture of the synovicytes, it reduces the activity of the genes responsible for synthesis of agrecanase-2 (the enzyme that destroys proteoglycans) and genes for synthesis of the inflammatory mediators TNF, IL-8, iNOS [68]. These effects are mediated through the receptors CD44 [69,70]. Contrary to bone, the cartilage is capable of interstitial growth due to the cells, which are surrounded by matrix. The cartilage cells (chondrocytes) synthesize hyaluronan, with which aggrecans are connected. As a result, the supramolecular complexes occupy a significant part of the extracellular matrix [71]. Each cell occupies a small cavity (lacuna) in the extracellular matrix. The cartilage has no blood vessels or even capillaries, so vital functions of chondrocytes are maintained by diffusion of the nutrients and gases from the blood vessels located far from the cartilage. Such diffusion takes place through the matrix and the synovial fluid before returning to the blood vessels. The cartilage is usually surrounded with perichondrium, the thick layer of the connective tissue that contains a significant amount of collagen and fibroblast-like cells. The perichondrium grows from inside as the chondrocytes secrete the matrix components. The fibrous perichondrium acts as a corset that prevents potential shape changes. As cartilage grows, the chondrocytes divide, resulting in the formation of new cells. The cells can enter the cartilage tissue from perichondrium. Then fibroblast-like perichondrium cells divide and undergo a transformation that results in the formation of the cartilage matrix. This is how these cells become chondrocytes.

Contrary to the cartilage matrix, the bone matrix is hard, which is why a bone can grow only by apposition, which is superposition of additional matrix and cells on the bone surface. An embryo creates a set of bone 'models'. As the new cartilage is created, the old one is replaced by the bone. The adult organism maintains a similar 'embryonic' form of the cells behaviour when the bone must be renewed. The bone is covered with the layer of thick connective tissue called periosteum. The bundles of collagen fibres emerging from the periosteum are inserted into a bone and provide a basis for attachment of other connective tissue such as tendons. The inner layer of periosteum includes osteoblasts, which are cells with an ability to proliferate. The bone matrix secrets these osteoblasts, which are located on the surface of the existing bone matrix, and layers new bone matrix over the top. Other cells including osteoclasts can destroy the cell matrix. Surrounding the bone there are cells capable of creating the cartilage matrix. At the break in the bone, these cells reproduce the initial embryonic process: first the cartilage is created in order to fill the gap and then the cartilage matrix is replaced by the bone matrix. The 'collaboration' of the cartilage and the bone is based on their common origin. Both tissues are formed from mesenchymal cells that secret large amounts of extracellular matrix with HA. Upon maturation of the cartilage, some of its parts are mineralized by the deposition of calcium phosphate crystals and other salts in the cartilage matrix. In these areas, chondrocytes undergo apoptosis, which leaves large voids. These lacunas are filled with blood vessels and osteoclasts, which destroy mineralized cartilage matrix. The following osteoblasts initiate a process to deposit the bone matrix. After this process, the only thing that remains from the cartilage in the long bones of the adult organism is the layer of cartilage tissue, which creates a smooth surface in the area of the joints where one bone is attached to another one. The role of HA in such process is not quite understood. It is clear that a bone, as a whole organism, is a dynamic system that maintains its structure due to a balance between opposite types of functions of the different specialized cells.

In the cases of arthritis or arthrosis, the irregularities of hyaluronan's function in the joints and pathologies are well documented. This is due to the fact that the osteoarthritis is a common disease associated with metabolic disorder, degenerative damages of the joint cartilage and adjacent bone tissue. Glenoid cavity and the head of the articulated bones are covered with a smooth and shiny hyaline cartilage that contains about 1 mg of HA per 1 g of raw tissue. The outer layer of the cartilage is coated with a shell that resembles periosteum and performs similar functions. The mating surfaces of the bones and articular cartilages are surrounded with the joint capsule. The inner surface of the capsule is lined with a membrane that secretes a viscous synovial fluid containing a significant amount of hyaluronan (3–4 mg/ml). It acts as a lubricant and performs spring functions.

Hyaline cartilage contains mainly extracellular matrix (95%) with a relatively small amount of chondrocytes that synthesize HA, chondroitin sulfate and other components of the extracellular matrix. Compared to fibroblasts, chondrocytes synthesize less HA, which is almost completely connected to aggrecans, to form the supramolecular complexes. Synthesis of chondroitin sulfate is carried out with the participation of six types of glycosyltransferases, which are specific for the six types of inter-saccharide bonds. Further esterification takes place with assistance of two sulfotransferase due to presence of two types of sulfate bonds in positions 4 and 6 of the amino sugars. Each chondroitin sulfate chain usually comprises at least 40 saccharide units and has a molecular weight of about 20 000 Da. Many such chains (about 100 chondroitin sulfate chains and about 30 keratan sulfate chains) are connected by covalent bonds with a small core protein. As a result, a macromolecule of proteoglycan aggrecan forms. This is the main protein of cartilage matrix (10% by weight). Proteoglycans are attached to HA by two proteins. Each hyaluronan macromolecule can be attached to 100 aggrecan molecules and form supramolecular complexes of the extracellular matrix. They bind with collagen structures and provide the strength, resilience and elasticity of the cartilage. Another major component of the extracellular matrix of cartilage is collagen type II [72].

Hyaluronan from the synovial liquid is produced by synoviocytes (a type of fibroblast). Synoviocytes assure slip of the joint surfaces and cartilage nutrition. The concentration of hyaluronan in the synovial liquid is quite high (2–4 mg/ml). Its molecular weight distribution is from 100 000 to 10 000 000 Da. Due to its high hydroscopic nature, it maintains water balance in the synovial liquid. The pathologies appear at certain conditions, including when synoviocytes start producing excess of cytokines, metalloproteases, hyaluronidases, eicosanoides (products of metabolism of arachidonic acid) and nitrogen oxide, and also when active forms of oxygen accelerate the degradation processes of HA and other biopolymers in the synovial liquid. The imbalance of synthesis and degradation of biopolymers towards acceleration of decomposition leads to the degradation of articular cartilage.

The upper layer of the articular matrix layer, which is in direct contact with the synovial fluid, is coated with a non-cellular plate with a thickness of 4–7 μm. The matrix of the cartilage surface layer mainly consists of the collagen fibres closely adjacent to each other. They are located parallel to the surface of the cartilage. Interfibrillar gaps in the surface layer are filled with the material, which consists of proteoglycans, hyaluronan and water. In response to the pressure, intercellular water is squeezed out of the cartilage plate into the joint space. After removal of the load, water comes back to restore the super hydrated cartilage state. Water, which is squeezed out of the cartilage, takes up most of the compression pressure and provides lubrication of the joint surface. Collagen content of the femoral head's cartilage surface layer is 8–16% higher than such content in the entire cartilage and glycosaminoglycans content is 1.6–2.0 times lower than in the entire cartilage. Consequently, the ratio of collagen to glycosaminoglycans in the surface layer of articular cartilage is 2–3 times higher than in the entire cartilage. However, articular cartilage surface layer is 8–10% more hydrated than cartilage tissue in general. It seems strange that at a reduced content of glycosaminoglycans, which are the most hydrophilic components, the surface layer of cartilage has a higher water holding capacity than the whole tissue. This contradiction could be explained by the fact that an increase in the ratio of collagen to glycosaminoglycans leads to a screening reduction of positively charged groups of collagen with negatively charged groups of glycosaminoglycans, which increases the degree of collagen hydration [56]. In collagen, at the maximum relative humidity almost all of the water content is associated with the active groups of the biopolymer. But in glycosaminoglycans, at least two forms of free water were found in addition to bound water [73]. It was experimentally established that at a low content of glycosaminoglycans, the surface layer of the articular cartilage has a higher water capacity compared to the rest of the tissue due to higher content of bound water. In the development of osteoarthritis, a change of the surface layer structure can be observed, which results in reduction of its thickness and change in structure of the fibre layers. In addition, the bound water content in the articular cartilage is reduced by approximately 10 times. In general, the water exchange balance in the joints depends on the HA condition in the synovial fluid.

According to modern understandings, the core causes of osteoarthritis are pathological changes that originate as a result of the disturbance of the normal processes in chondrocytes and degradation of structures in the extracellular matrix of the articular cartilage [58,59]. During the development of such pathology, possibly due to genetic reasons, a decrease in size of hyaluronan molecules as well as other glycosaminoglycans is observed, which means that biomolecule depolymerization has taken place (these processes are discussed in detail in Section 4.3). Therefore glycosaminoglycans, which can be found in cartilage, are represented as smaller subunits that could be easier eliminated from the matrix and migrate outside the joint capsule. Such changes are already recorded at the early stages of osteoarthritis and their intensity is correlated with the development of the process.

In the chronic course of the disease, the degraded fragments of glycosaminoglycans in complex with proteins may cause antigenic stimulation and result in an increased level of antibodies sent to complexes of hyaluronan and other complexes of glycosaminoglycans sent to proteins. At high levels of antibodies in these complexes a significant reduction in the number of positive results in the treatment of the joints diseases with chondroprotective agents (chondroitin sulfate, glucosamine, etc.) was observed [74].

The development of osteoarthritis often begins with cytokines, which are growth factors synthesized in the joint tissues, then transferred into the synovial fluid where they act as autocrine and paracrine factors for the maintenance of homeostasis. Osteoarthritis does not show the classical macroscopic signs of inflammation revealed in infiltration of joint cells and tissues. However, the increased amounts of anti-inflammatory cytokines such as interleukin-1-beta, tumour necrosis factor-alpha can be found in the patient's synovial fluid. Under the influence of interleukin-1-beta, chondrocytes sharply increase the synthesis of metalloproteinases and stop the synthesis of proteoglycans and cartilage collagen. Under normal conditions, chondrocytes synthesize and export into extracellular matrix proteolytic enzymes that include metalloproteinases, cysteine and serine proteases. The controlled activity of these enzymes is needed for the permanent exchange-remodelling process of the extracellular matrix cartilage [56–59].

The products for the treatment of osteoarthritis could be divided into two groups: those that modify the symptoms and the products and those that modify the structure. The second group includes chondro-protectors, which have the ability to restore the structure of the cartilage tissue. These chondro-protectors are the products of cartilage tissue depolymerization (glucosamine, glucosamine sulfate and chondroitin sulfate) and HA, which is needed for restoration of viscoelasticity of synovial fluid in the joint capsule. The chondro-protectors (Table 6.4) are considered to be a basis of the constructive treatment of joints diseases. The exogenous glucosamine passes through the gastrointestinal tract until it reaches the cartilage where the cells attach it to HA, heparin and chondroitin sulfate. In the body, chondroitin sulfate can be found mainly in the cartilage matrix where, together with hyaluronan, it retains water and thus contributes to the elasticity and shock-absorbing tissue properties. In addition to that, chondroitin sulfate is an inhibitor of metalloproteinases and has anti-inflammatory, analgesic and anti-atherosclerotic effects. Sulfated glycosaminoglycans such as chondroitin sulfate and heparin showed a dose-dependent ability to inhibit the activity of aggrecanases in cell cultures [58]. Glucosamine and chonroitin sulfate help to restore the structure of cartilage tissue and act as synergists, enhancing their mutual action. Mechanisms of chondro-protective action are directed to stimulate the synthesis of cartilage matrix components, slow its destruction and activate the metabolic activity of the cell. It was found that chondrocytes are not only capable of synthesizing the precursors of glycosaminoglycans but also of utilization the molecules that arrive in the medicinal products. The finished macromolecules are used in the extracellular matrix during self-assembly of its structures [57].

The second approach of biomedical technologies for the treatment of osteoarthritis is related to hyaluronan. Patients with deforming arthritis have a reduced concentration and molecular weight of HA in the synovial fluid of the joint compared to the norm. This leads to a deterioration of the viscoelastic properties of the biopolymer and synovial fluid in general. A reduced amount of hyaluronan leads to a reduced ability to protect joint surfaces from damage and wear and joint pain receptors from unnecessary irritation. A simple injection of exogenous high molecular weight HA into the joint cavity leads to restoration of the viscoelastic properties of synovial fluid. It also leads to stimulation of the synthesis of polysaccharides by synoviocytes and synthesis of cartilage proteoglycans by chondrocytes. Proteoglycans, in turn, prevent the destruction of the chondrocytes and inhibit the release of the enzymes that can destroy cartilage. Furthermore, HA has an analgesic effect due to its ability to block the joint pain receptors. It is assumed that hyaluronan also reveals anti-inflammatory activity by possessing the ability to inactivate free radicals, inhibit leukocytes migration and chemotaxis. The drug, based on hyaluronan, has been recently introduced on the market under the name *Gialgan*. It is a purified solution of the sodium salt of hyaluronic acid with a molecular weight of 500 000–700 000 Da.

The size of Gialgan macromolecules is significantly lower than HA in normal synovial fluid. Therefore, when injected into the joint capsule, Gialgan increased synovial fluid elasticity by only 1–15%. As a result of hyaluronan depolymerization in the joint, the duration of the action of such temporary 'prosthesis' is limited to 2 days. By chemically cross-linking of HA molecules, the new product *Synvisc* was obtained. Its molecular weight is about 6 000 000 Da. Due to intermolecular cross-linking and an increase of the molecular weight of the biopolymer, the synovial fluid elasticity was increased up to 78% and the action duration up to 7 days. A number of the experimental studies were conducted to elucidate the efficiency and mechanisms of action of various drugs based on hyaluronan, such as Gialan, Synvisc, Restylane, Dermalife and others, which should be injected directly into is joint capsule of knee (gonarthrosis) and hip (coxarthrosis). The overall conclusion is that HA provides a positive effect on cellular and immunological functions, especially in the early stages and on moderate diseased symptoms. A direct relationship is established of the various biological actions on the HA molecular weight – the larger the molecular weight is related to better and longer effect. It is shown that HA with a molecular weight of 500 000–750 000 Da in combination with chondroitin sulfate in a wide range of concentrations (10–1000 mcg/ml) inhibits the synthesis of stromelysin-1 (metalloproteinase-3, which destroys cartilage and acts as an inflammation mediator).

It is proved that various changes in the metabolism of proteoglycans in the articular cartilage and synovial fluid in the development of degenerative and dystrophic processes in the knee are interrelated and interdependent [53,56]. The different analyses and studies of the synovial fluid allow for a better understanding of the ratio of the processes of degradation and regeneration of articular cartilage.

6.3   Hyaluronan in Ophthalmology

Hyaluronan based products are widely used both in the general and local therapy of eye diseases. Hyaluronan is included in the composition of eye drops ('artificial tears') for the treatment of dry corneas. There is a drug called 'Healon' developed by Pharmacia, Sweden that contains a high molecular weight HA and is widely used for eye surgery procedures and for surgical media (viscoprotector) to prevent the eye's internal tissues from mechanical damage.

The vitreous body of the eye largely consists of HA produced by the cornea cells. Due to the viscoelastic properties of hyaluronan, the vitreous body assures permanent intraocular pressure and prevents detachment or retina and pigment epithelium.

Several formulations of HA solutions with amino acids, antioxidants and other compounds have been developed. They can be prescribed for eyeball irrigation, or injection into the eye's anterior chamber during cataract surgery [61,75,76]. The products of the Hyalrepair line could be very useful for such purposes as well.

6.4   Hyaluronan in Oncology

Hyaluronan is used in oncology as a therapeutic. There are various mechanisms of HA action on tumour cells. At the molecular level, the main mechanism is related to the fact that the high molecular weight HA can bind to receptors on the cell membrane of tumour cells, which results in slowed cell migration and metastasis [77]. For cancer cell invasion and metastasis formation, the proteolysis of the extracellular domain of the receptor CD44 is required. In this case, the cell is released from its connection with hyaluronan and is removed from the 'anchor' [78,79]. Thus, the high molecular weight polysaccharide can be used as an inhibitor of cancer growth and metastasis formation [61]. Another mechanism of action is related to the ability of high molecular weight HA to promote the formation of a capsule surrounding the tumour. The capsule is built from the connective tissue. A third mechanism is related to the ability of high molecular weight HA to inhibit tumour vascularization (sprouting of blood vessels in the tumour), which leads to the slow growth and metastasis of tumours [80,81]. The sprouting of the new blood vessels in the tumour (angiogenesis) is a key element of its progression. Fibroblast growth factors secreted by tumour cells stimulate proliferation of endothelial cells and promote the formation of new capillaries. Angiogenesis provides cancer cells additional benefits of growth and invasion. In this context, the researchers are faced with the duality of the physiological effect of the hyaluronan. Contrary to high molecular weight HA, which inhibits angiogenesis, the low molecular weight fraction (below 30 000 Da), induced it [82]. For invasion, the tumour cells secrete extracellular hyaluronidase, which hydrolyses the hyaluronan of the tissue extracellular matrix. Therefore, there is a need for drugs (and possibly complexes) of high molecular weight HA in combination with potent hyaluronidase inhibitors.

Another remarkable property of hyaluronan is that it enhances the effect of conventional anti-cancer drugs. HA allows their therapeutic doses to be reduced and consequently adverse toxicity to normal cells [83]. A large amount of hyaluronan and hyaluronidases is found in the extracellular matrix of the tumour. Many studies on the large number of various types of tumour cells demonstrated that high molecular weight HA can inhibit both proliferation and migration of tumour cells [80,81,83].

It is assumed that the tumour cells' use of hyaluronidase as a 'molecular sabotage' for the depolymerization of extracellular matrix polysaccharide macromolecules for two reasons: first, to facilitate the invasion between normal cells and form metastasis in the tissues and secondly, to use the ability of low molecular weight degradation HA products to stimulate angiogenesis by sprouting blood vessels in tumour tissue. In order to discuss the molecular mechanisms by which hyaluronan affects the cell proliferation and tumour growth, it is important to examine the differences between normal and tumour cells [72].

Tumours represent a group of genetic diseases characterized by uncontrolled cell proliferation. The tumour cells typically are round or star-shaped and they are larger than normal cells. They have a different nucleus-cytoplasm ratio. The nucleus often contains not diploid, but polyploid or aneuploid sets of chromosomes. The differentiated cells are kept within certain tissues and do not intrude into the 'adjacent areas'. They have a number of advantages over normal cells because of several characteristic features:

1. Increased rate of DNA and RNA synthesis by increased activity of the enzyme ribonucleotide reductase and a reduced rate of catabolism of purines and pyrimidines.
2. Accelerated rate of glycolysis and increased lactate production. The activity ratio of key enzymes of glycolysis and respiration is changed and correlates with tumour growth rate [84].
3. Protein and enzyme content is increased. So, the isozyme of tumour hexokinase has an extremely high affinity for glucose. Such changes in the carbohydrate metabolism provide cancer cells with the ability to assimilate glucose, even at very low blood concentrations. It gives another advantage to the tumour cells in competition with the normal cells. Similar shifts could be observed and found in other processes that allow tumour cells to successfully compete with their surrounding tissues for vital metabolites.

4. Embryonic proteins and enzymes could be found in the tumour cells. In particular the highly active enzyme telomerase, which is characteristic for embryonic cells, appears in the tumour cells. The main function of telomerase is to add telomeres to 3' end of one DNA chromosome strand. Telomeres represent the end fragments of the linear chromosomes of human and advanced animals. There are thousands of highly conservative repetitions of the hexadeoxynucleotide TTAGGG that are named 'telomeres'. The telomere parts of the chromosome are attached to the nuclear membrane and, as a result, prevent destruction and recombination of chromosomes in the differentiated non-proliferating cells. At each cell cycle and DNA replication, the length of telomeres is shortened by approximately 120 pairs of the nucleic bases, which leads to the inability of DNA polymerase to add end fragments at DNA 3' end. The normally divided somatic cells do not have telomerase. Therefore, the shortening of the telomeres serves to measure the number of the replicas that determine the number of divisions (Hayflic limit) and the lifetime of the normal cell. After reaching the critical size of telomeric sequences, the cells lose their ability to divide and undergo programmed cell death (apoptosis). In embryonic and tumour cells, telomerase is active and completes the telomeres at the ends of the 3' end DNA after replication and restores their original length. As a result, the cell's ability to age is terminated and they become immortal.

5. The transformed cells have the modified composition and structure of the oligosaccharide chains of glycoproteins and glycosphingolipids of plasma membrane. As a result, the structure, charge and membrane permeability are changed and the intensity of the synthesis of adhesive proteins and integrin receptors is reduced. The proteases, collagenase and hyaluronidases are secreted into the extracellular matrix. They hydrolyse collagen, hyaluronan and other glycosaminoglycans decompose the structure of the extracellular matrix and facilitate the invasion of tumour cells into the tissue. The synthesis of angiogenesis factors, which contribute to blood vessel sprouting and supply the tumour with its metabolites, is accelerated.

6. In both normal and tumour cells, growth and cell division begins from the action of the growth factors on the cell. By interacting with receptors on the cell surface or intracellular receptors, they turn on the cascade of reactions inside the cell. This leads to the activation of genes responsible for the synthesis of proteins of cell division. If the genes encoding the receptors, transducers and transcription factors are modified as a result of mutation in such a way that the expression is not coordinated, then the controlled cell division is replaced by non-controlled cell division immortalization (cell transformation). In tumour cells the rate of synthesis and secretion of certain hormones and growth factors is increased. The cells acquire the ability for autonomous growth by shifting to paracrine or autocrine mechanisms of growth regulation. The autocrine regulating mechanism allows tumour cells to synthesize growth factors, receptors and oncoproteins, which are the analogues of growth factors or their receptors. Interacting with each other, they cause autostimulation of the cells' growth and division.

7. The activity of proto-oncogenes and tumour suppressor genes changes. Most tumours are originated from somatic cells. Because somatic cells are diploid, they have two alleles of each gene. If the mutation(s) in one allele lead(s) to dysfunction of the daughter cells, it means the dominant type of inheritance characteristic for proto-oncogenes and the gene P53. Proto-oncogenes control the synthesis of proteins, which trigger the cell cycle and thus may reveal the oncogenic properties. Such genes are acquired from

a number of human and animal viruses. In order to differentiate between normal genes and viral oncogenes, the definition for the host genes is proto-oncogenes. This group includes the genes that control the synthesis of certain proteins. These proteins play a central role in the regulation of cell division such as growth factors, growth factor receptors, transcription factors and proteins involved in signal transduction. Half of the viral oncogenes encode tyrosine proteokinases. Proto-oncogenes contain information about the family of Ras-proteins that possess GTPase activity and participate in signal transduction. Obtained in cell membrane receptors and localized at the inner side of the membrane, they are in the close contact with membrane proteins and phospholipids. It was shown that Ras-proteins are involved in the change of cytoskeleton structure, regulation of exocytosis and endocytosis, implementation of the mitogenic signals and activation of proteins and the regulation of gene transcription. There is a special family of nuclear proto-oncogenes; a group that includes genes such as myc, fos, jun, myb and erb A. Oncoproteins produced by these genes are bound with specific sequences on DNA and function as transcription factors.

The antioncogenes or tumour suppressor genes are described as well. The protein products of these genes inhibit the cells' replicative potential and prevent tumour development. In the course of malignant transformation, the functions of these genes are often lost, which leads to impaired control of cell proliferation. In this regard, the best studied gene is gene P53. It encodes a nuclear phosphoprotein with a molecular weight of 53 000 Da that prevents cells from moving into S-phase as well as amplification and gene mutation.

It is believed that the physiological function of the protein P53 is mainly to delay the cells (which have damage in DNA structure) in G1–G2-phases of the cycle of cell division until these damages are eliminated by the damage repair systems. If reparative systems cannot eliminate defects in DNA structures, the protein P53 initiate apoptosis mechanisms that destroy the damaged cell. The P53 protein tetramer binds to the regulatory areas of many genes and modulates their expression. By activating key genes, which initiate apoptosis, the protein P53 accelerates the destruction of potentially dangerous cells that are damaged and capable of transformation. The protein P53 increases expression of the gene encoding the protein thrombospondin. This protein prevents sprouting of the blood vessels into tumour and metastasis formation. Thus, the protein P53 functions as a 'molecular policeman', maintaining healthy cells and eliminating the damaged ones. During the human life, there are totally about 10^{16} cell divisions, during which each gene can mutate 10^{10} cases. Several scientists believe that the problem of cancer is not why it occurs, but rather why it occurs less frequently. There are cell reparative systems and gene suppressors, which in most cases eliminate the unlimited reproduction of differentiated cells.

Benign tumours can grow rapidly and reach large sizes but do not metastasize. Only malignant tumours can invade the other tissues (metastasize). When a cell tumour reaches about 2 mm in diameter, the cells secrete protein factors that stimulate the growth of connective tissue around the tumour and the growth of blood vessels into the tumour. Angiogenesis provides tumour cells with additional benefits for growth and invasion. Metastatic cells secrete a range of enzymes for destruction of extracellular matrix and basal membranes of the surrounding tissue. These enzymes include collagenases that can cleave collagen of the extracellular matrix and a powerful protease cathepsin B that in normal cells is localized in lysosomes and in metastatic cells is embedded into the plasma membrane.

Cathepsin B activates procollagenase, which specifically cleaves type IV collagen. The list of enzymes also includes plasmin (the enzyme, which breaks some non-collagenous proteins of the extracellular matrix), the family of metalloproteinases and heparase (the enzyme that catalyses the hydrolysis of heparan sulfate: the predominant proteoglycan of the basal membranes). The fragments of the extracellular matrix, including HA surrounding tumour cells, camouflage these enzymes from immunological surveillance and provide attachment to the basal membrane in the target organs. Carbohydrates located on a tumour cell surface are binding to the carbohydrate component of the endothelial cell receptor and finally attached to the blood vessel walls with help of the integrin protein.

Most differentiated cell populations of human and vertebrate animals undergo rejuvenation. Tissue integrity is maintained by differentiated somatic cells that sometimes leave the differentiated state and enter into the mitotic division cycle to replace damaged cells. This route (differentiated state –>cell cycle initiation –>mitotic cycle –>mitotic cycle termination and return to the differentiated state) has a substantial risk of mutagenic, carcinogenic and other external factors, failure in the regulatory systems and linkages that could lead to unlimited cell proliferation (immortalization), transformation, malignancy and the cancer process, which cause death for the whole body.

According to WHO (World Health Organization), 80% of human cancer diseases are induced by environmental factors. The carcinogen dosages in the environment are low, but they act in combination with other factors that induce the increased proliferation of differentiated cells and therefore create the necessary conditions for transforming carcinogens' actions. It is difficult to predict the outcome of the complex influence of physical and chemical factors because of a large variety of incoming signals as well as insufficient knowledge of the behaviour and interaction of various cells systems in such conditions. The combined attack on the cells is characterized by such phenomena as competition, additive and synergistic effects, collateral sensitivity or cross-resistance. In extreme or even non-optimal conditions the differentiated cells can move into the states of proliferation, transformation, stability, apoptosis or chain cytolytic processes.

Based on the hypothesis that the extracellular matrix supports the cell phenotype (which synthesizes this matrix), it is possible to hypothesize that the transition state selection of the cell is initiated from the extracellular matrix since it is the first to receive the signals from the extracellular media and is exposed to external physical and chemical factors. The structures of the extracellular matrix and first of all hyaluronan and proteoglycans can protect the cells from external factors but also accumulate mutagens and carcinogens. These two factors can change the synthesis-to-breakdown ratio of the extracellular matrix components, and therefore remove the cells from the differentiated state and initiate transition states. This is confirmed by the following data: when typical cartilage matrix proteoglycans are added to chondrocytes, matrix component synthesis is activated in the chondrocytes. Free HA, which is added to the culture of chondrocytes, strongly inhibits the synthesis of cartilage matrix [85]. These results suggest that the extracellular matrix, which is secreted by the cells, affects the cells and helps them to maintain a specific differentiated status. The excessive above-threshold effect helps to move the cell into the mitotic cycle. According to our data, the transition of the differentiated cells into the cell cycle is associated with major rearrangements of chromatin in the nucleus. The transition is accompanied by several metabolic changes and activation of nuclear proto-oncogenes that were 'silent' in the differentiated cells. The result is the suppression of activity of 'differentiation genes'; activation of the group of 'proliferation genes'. Ultimately, the cells switch from synthesis of secreted biopolymers to the synthesis of intracellular proteins on the basis of the competitive relationships of proliferation and differentiation. Based on these data, general principles of differentiated cells transition in the cell cycle have been formulated [86]. This major reorganization of chromatin in the nucleus and the cell cycle in general apparently occurs with the cell and HA, is localized in the nucleus and delegated from the cellular matrix. However, these questions remain to be studied. At present it is known that the synthesis and secretion of hyaluronan increases during proliferation. In the mitotic cells, this polysaccharide fills the entire area around the chromosomes during metaphase, anaphase and telophase [87]. The tetrasaccharide HA can regulate the expression of genes for certain heat-shock proteins [88] that, like the nuclear proto-oncogenes, belong to the 'early response' group of genes. Studies exist that explore the signal transduction pathways mediated by hyaluronan. Thus, during endothelial cell mitogenesis, which is caused hyaluronan oligosaccharides (oHA), the activity of the protein kinase C and MAP- kinases is stimulated and 'early response' genes are activated in endothelial cells along with the synthesis of Ras- protein. The 'early response' genes include the previously mentioned nuclear genes *myc*, *fos* and *jun* as well as few others. It is shown that exogenous HA inhibits migration of fibroblasts transformed by the gene *ras*, a protein fragment that specifically binds a polysaccharide on the surface of fibroblasts. Also, the same protein fragment reduces the ability of the polysaccharide to inhibit locomotion of *ras*-transformed fibroblasts. The specificity of this action could be understood by the dose-dependent effect and the fact that the chondroitin sulfate and heparan sulfate do not show such effect [77].

Antiproliferative activity of hyaluronan with a molecular weight of 500 000–750 000 Da at concentrations of 0.8–80 mcg/ml was tested *in vitro* on a number of cancer cells [83]. It was found that HA suppresses the cells proliferation of prostate cancer cells, bladder cancer, melanoma, fibrosarcoma and breast cancer. The degree of suppression depends on the biopolymer concentration. The same paper contained the analysis of the possible mechanisms of antiproliferative activity of hyaluronan for cancer cells. Inhibition of cancer cell proliferation by hyaluronan was not dependent on the level of receptor CD44. The glycoprotein CD44 is associated with a malignant process in several types of cancer and contributes to survival of cancer cells. It was reported that the suppression of CD44 in breast carcinoma induced apoptosis in cancer cells [89]. So far no correlation was described between the level of expression and the number of receptors CD44 on the surface of cancer cells from one side and the antiproliferative activity of the HA from another side. However, there are several data that suggest that such correlation could exist. For example, anti-proliferative activity of HA on the cancer cells can be explained by the destruction of the complexes of the polysaccharide with receptor CD44 [90]. Another reason for anti-proliferative activity of HA could be related to another receptor, for example RHAMM. The isoforms of the receptor RHAMM are involved into regulation of the cell cycle as well.

It was found that RNAMM overexpression by transfection into normal fibroblasts turns them into metastatic fibrosarcoma cells. The ability of hyaluronan to inhibit cancer cell growth is not related to mutations of r53/r21 or Rb-, deletion of p16, Fas-stability, absence of caspase-3 or P-glycoprotein overexpression.

It is interesting that the anti-proliferative action toward tumour cells could be shown by hyaluronan with a wide range of the molecular weight. *In vitro* experiments demonstrated that at concentrations greater than 0.32 mg/ml, hyaluronan fractions with a low molecular weight (below 5600 Da), and with high molecular weight (above 1 200 000 Da), both inhibited melanoma cell proliferation up to 50–90%. These are the cells of highly malignant rat melanoma, which express the receptor CD44. Other controversial data was presented that explores the role of hyaluronan in the proliferation of the normal cells. It was found that HA with a molecular weight of 1300–4500 Da (3–10 disaccharide units) causes proliferation of endothelial cells. This phenomenon was correlated with an increase of the content of protein kinase.

On the contrary, a polysaccharide with a molecular weight in the range of 400 000–1 000 000 Da inhibits the proliferation of normal endothelial cells. In a number of studies it was shown that hyaluronan modulates cell proliferation in various types of normal cells. The inconsistency of the results, which could be found quite often in the analysis of the published data regarding the HA effect on the different cellular processes, may be related to differences in the expression of hyaluronan receptors, the presence of impurities, the molecular weight dispersion and source of this polysaccharide. Nevertheless, it is possible to say that HA with a molecular weight of 500 000–750 000 Da enhances the activity of such anti-cancer agents such as 5-fluorouracil, cisplatin and tamoxifen.

It was also found that hyaluronan acts both as an additive and synergist with anti-cancer drugs [83]. This insinuates that hyaluronan has significant potential in the realm of anti-cancer drug development as a chemotherapeutic agent, an additive to anti-cancer agents and as a booster of anti-cancer drugs. Additional properties of HA could be seen in complexes with several metal cations (zinc, copper, silver, gold, etc.). Similar activity was demonstrated for hyaluronan in composition with antiviral, immune-stimulating and anti-bacterial agents. It was shown that hyaluronan increases the synthesis of interferon by T-cells and monocytes, which are the main interferon inducers [91,92].

Composition of HA with RNA – interferon inducer (0.15% RNA and 1% HA) is the greatest interferon activator in the animals. The mechanism of action is likely related to the fact that the polysaccharide macromolecule blocks some molecular inflammatory factors such as CD44 and RNAMM [93,94], has some anti-tumour properties [81] and activates the metabolism and proliferation of normal cells. It is also a possible direct effect of biopolymer on the synthesis of interferon by lymphocytes and monocytes through activation of CD44 in these cells [91].

Many functions have been described for the main protein receptor CD44, which binds HA from the extracellular matrix with the cell surface [95,96]. The trans-membrane glycoprotein CD44 contains ectodomain on the surface of the cytoplasmic membrane, another domain that is localized in membrane and the intracellular domain. Ectodomain binds HA to the extracellular matrix, a process that occurs with more than one molecule of CD44 and when length of the binding part of the macromolecule is more than 20 monomers. It is found that the ectodomain of CD44 can be proteolytically cleaved by membrane-bound metalloproteinases.

Cleavage of the ectodomain CD44 leads to the release of cells that have been bound with hyaluronan, which is how cell migration is regulated. Inhibition of cleavage leads to the inhibition of tumour cells that migrate on a substrate of HA. It is possible that fragmentation of hyaluronan associated with hyaluronidase CD44 secreted by cancer cells opens paths of migration and metastasis of cancer cells. Exogenously introduced biopolymers can inhibit these processes. An increase of the lung metastases after injection of carcinoma cells was confirmed experimentally. This phenomenon is related with activity of matrix metalloproteinase-9, which promotes the degradation of collagen IV and thus mediates an invasion of tumour cells.

The CD44 receptor can bind metalloproteinase-9 outside a hyaluronan-binding motif in areas sensitive to hyaluronidase and recruit it on the membrane (platform function). It cleaves the ectodomain CD44 only after binding to the ectodomain of HA and internalization has occurred. It was noted that in tumour cells, which are able to bind HA with CD44, cell migration is slowed and metastatic growth is reduced. In such a situation, the binding of HA with CD44 likely prevents apoptosis of tumour cells. It is believed that CD44 plays a critical role in the metastasis of tumour cells but not in the initiation of carcinogenesis [90,96].

6.5   The Role of Hyaluronan in Healing Wounds

Hyaluronan has a number of properties that favourably distinguish it from many other products used for healing wounds. Namely, HA has no allergenic or irritating actions but can provide both anti-inflammatory and bio-stimulating effects and accelerate the regeneration processes. These properties of HA, in combination with other drugs, are used to accelerate the healing of burns, trophic ulcers and surgical procedures. The healing of skin wounds is directly related to an increase of HA during the first three days and a decrease to the initial level by seventh day after injury. In the surgical dressing the polysaccharide is used both as a primary wound healing product and in combination with other agents.

The film on the base of the oxidized HA demonstrated ability to accelerate healing of suture (of the intestinal high-risk anastomoses. It can also be used to close the defective serous membrane and for prophylaxis of the bowel perforation during laparoscopy [97,98]. Hyaluronan is used for the treatment of the stomach and duodenum ulcers. The anti-ulcer action of HA is related to its ability to inhibit H2 histamine receptors and the activity of trypsin [61]. Due to the ability of low molecular weight fragments of HA to penetrate the skin epidermal barrier, it is used as a transporter of bioactive compounds. According to some sources, only HA fragments with a molecular mass of 400 kDa can penetrate through the skin [99]. The smaller the fragments of the polysaccharide macromolecule the higher the permeation rate [76,99,100].

As mentioned earlier, hyaluronan is a high-rate turnover biopolymer. Its half-life in the skin is about 24 h. Any changes in the processes of hyaluronan turnover (rate of the ratio of synthesis:decomposition) could cause development of pathologies. When tissue is damaged, the existing hyaluronan undergoes decomposition, resulting in appearance of oligosaccharide fragments that are involved in the formation of a temporary extracellular matrix. At the same time, synthesis and accumulation of hyaluronan is rapidly induced in the tissue damage area by activating genes and different hyaluronate synthases. This suggests that the macromolecule is an active participant of reparative processes of damaged tissue.

Wound healing is a dynamic process consisting of three overlapping phases: inflammation, proliferation and maturation (restructuring).

1. At the first stage (inflammation stage), the main process is the formation of a blood clot in order to stop bleeding. The blood clot is a temporary matrix consisting of proteins such as fibrin and blood thrombocytes. The clots are formed from blood as a result of damaged blood vessels. The inflammatory cells migrate into the clot. During the following

decomposition of the thrombocytes, several different growth factors are released, but at the decomposition of fibrin by proteases, a number of signal molecules (polypeptides) appear. The polypeptides attract neutrophiles, monocytes and fibroblasts. Even at the early stages, an increase of the concentration of hyaluronan is noticed, which binds to the fibrin network, forming a transition matrix. It interacts with cells by using receptors and a large number of proteins, namely proteoglycan to form dense 'sieves'. Its dispersing matrix forms small tubules for the selective diffusion of water-soluble molecules. The macromolecule of HA contains regularly repeating hydrophobic regions, which allow for interaction between cell membranes and matrix structures. At the same time, the enzymes of the extracellular matrix are activated. These enzymes catalyse decomposition of proteins and polysaccharide polymers to the fragments of lower molecular weight and new functions. These protein and polysaccharide fragments participate in the formation of the transition matrix. The structural and functional properties of the different forms of transition matrix seem to be different. It is possible to see two main phases of the transition matrix: gel and liquid crystals. The high molecular weight HA, when inserted into the protein network, creates a gel-like matrix. Fragmented HA and polypeptides are able to form structures similar to liquid crystals. Liquid crystal structures have much greater mobility, which facilitates cell migration into a wound. In response to action of growth factors and signalling systems, a migration and proliferation of fibroblasts of derma adjacent to the wound is started. In an adult organism, the fibroblast proliferation precedes the synthesis of collagen (see Chapter 2).

2. At the second phase (proliferation stage or fibroblast stage or granulation tissue stage), the active synthesis and decomposition of hyaluronan and proteoglycans continues. This stage proceeds with active participation of metalloproteases, a group of matrix enzymes whose activity depends on zinc. The main part of the body zinc is located in skin. Metalloproteases catalyse the decomposition of the molecules of the extracellular matrix. Their activation mechanisms in the extracellular matrix are unknown. It is believed that in few cases the activation of enzymes occurs through autocatalysis. As a result, the N-terminus protease polypeptide is removed. Simultaneously the cells secrete metalloprotease inhibitors. The depth of the destruction of the structures of the extracellular matrix and creation of the temporary matrix depends on the interaction of the activators and inhibitors. Ultimately, it affects the remodelling of the damaged tissue. The second stage of the wound healing is characterized by active fibroblast proliferation, synthesis of large amount of hyaluronan and new collagen. The fibroblasts of the adult collagen predominately synthesize type I collagen. Interestingly, the embryonic fibroblasts in the early stages of foetal development produce more III and IV types of collagen. Similarly, the content of type III collagen is higher than type I collagen in the healing wounds. The rationale could be the following. Many growth factors and new signal fragments, formed at the decomposition of HA and proteins, initiate migration of stem (mesenchymal) cells into a wound. Inside the wound they differentiate into 'young' fibroblasts that are likely not limited by Hayflick number, divide fast and produce a significant amount of type III and IV collagen similar to embryonic fibroblasts. At the same time, a new extracellular matrix, which is closer to the structural and functional organization to foetal matrix, is formed. An important role in the healing of skin wounds is the basal membrane, which creates a platform for cell migration and accelerates a proliferation of the epithelial cells (wound epithelialization). Skin wounds change the

chemical composition of the basal membrane. For example, the amount of fibronectin is increased, which promotes cell migration similarly to HA.

3. At the third stage (maturation, ageing and reconstruction or reorganization), when granules form and an excess of collagen matrix appears, the number of cells is reduced by apoptosis [101]. The excess of hyaluronan undergoes depolymerization and the short fragments of the polysaccharide macromolecules stimulate angiogenesis (a sprouting of blood vessels in the extracellular matrix [82]). The temporary matrix is destroyed by the matrix hyaluronidases and metalloproteases in a process that is, essentially, a remodelling of the temporary matrix. Hyaluronidase not only hydrolyses HA but also cleaves chondroitin sulfate from proteoglycans. In the free state, sulfated glycosaminoglycans exhibit unique properties – they reduce the activity of hyaluronidases and proteolytic enzymes in the extracellular matrix, block synthesis of inflammation mediators by masking antigenic determinants and cancellation of chemotaxis, decrease the activity of free radicals, prevent cells apoptosis, inhibit the synthesis of lipids and prevent degradation processes in would healing. Interferons, which are produced by T-lymphocytes, leucocytes and fibroblasts, reduce the rate of synthesis of collagen. The same function is likely attributed to hyaluronan [102]. As a result of such a verity of process, equilibrium is reached.

It could be concluded that the processes of tissue and organ regeneration pass through stages of stimulation of the differentiated cells to proliferation, migration and again to differentiated stage. Hyaluronan of different molecular weights is important for each stage of the cycle of remodelling. Its role is not limited to moisturizing the wound surface. HA participates in remodelling of the extracellular matrix, the processes of cells differentiation, proliferation and migration, and establishment of new aqueous and ionic balances. That is why in the medical practice, one of the main applications of hyaluronan is acceleration of wounds healing, especially for burn wounds [102]. HA is extremely efficient for reconstruction of damaged epidermis. It was found that granular layer and the inner layer contain more hyaluronan than the upper skin layers. High molecular weight polysaccharide macromolecules are unable to penetrate the upper horny (keratinous) layer of the epidermis. Therefore, its role in topical application in solution or cosmetics is limited to formation of the molecular web that binds atmospheric moisture and reduces evaporation through the stratum corneum of the epidermis, thus maintaining skin hydration. Experimental studies showed that after the destruction of the epidermal barrier with acetone, HA accumulates in the epidermis between the basal and stratum spinosum (spinous layer) [103]. In response to injury, the epidermal becomes thicker than usual. This is an adaptive response that leads to reduction in further damage. Thus, hyperplasia of epidermis by UV irradiation reduces the damaging effect of radiation. However, for diseases such as eczema and psoriasis, hyperplasia is part of the pathological process. Management of epidermal processes could be a new approach for the treatment of these diseases.

The basal membrane has an important role in the regeneration of the wounds of skin and other tissues and organs. It creates a platform for migrating cells and changes the architectonic of the damaged tissue. In the skin basal membrane, the chemical composition is changed by the addition of fibronectin, which, as hyaluronan, accelerates cells migration.

Our generalized understanding of skin regeneration still leaves questions for discussion. At the same time it is evident that human skin is one of the indicators of health and HA is one of the main factors for its preservation and healthy condition. The experimental studies in such a direction have already resulted in a number of important products [11] and it is highly likely they will lead to creation of new therapeutic and cosmetic compositions based on hyaluronan [104].

6.6   Hyaluronan in Immunology

The immune system is connected with HA [91–94]. Hyaluronan is included into drugs used for the complex treatment of immunodeficient conditions associated with viral diseases. At the molecular level, the mechanism of action of the biopolymer is connected with the blocking of several molecular inflammation factors [93]. On the one hand hyaluronan activates inferonogenesis but on the other it increases action of the interferon inductor (e.g. double-stranded RNA) [91,92]. Interferon is produced mainly by the activated monocytes and T-cells of the immune system. The interleukins-2 and -5 (IL-2, IK-5) play a major role in the activation of T-cells that, in turn, activate synthesis of hyaluronan by endothelial capillary cells. Then hyaluronan stimulates synthesis of CD44 receptors, which is the key event for the activation of the lymphocytes and monocytes [93]. HA is used alone or in combination with interferon to slow down the development of the infection by the virus herpes simplex by application on the infected epithelium [61]. The obvious antimicrobial action of HA can be achieved by its cross-linking with hydrophilic polymers, which are capable of accelerated penetration through cell membranes or intercellular gaps [61].

Recently, several other applications of hyaluronic acid in pharmaceutical science and practice have been considered and are highly prospective, such as the application of HA as a drug delivery system to increase drug efficacy, prolong their action, reduce side effects and so on. However, this is very wide topic that could be addressed in the separate monograph.

References

[1] Edsman, K., Nord, L.I., Öhrlund Å., et al. (2012) Gel properties of hyaluronic acid dermal fillers. *Dermatologic Surgery, 38*, 1170–1179.

[2] Lopatin, V.V., Askadskii, A.A. (2004) *Polyacrylamide Gels in Medicine* (in Russian). Nauchny mir, Moscow, p. 264. Лопатин, В.В., Аскадский, А.А. Полиакриламидные гели в медицине. М.: Научный мир.-2004.-С. 264.

[3] Zhang, L., Zhang, W., Zhang, Z., et al. (1992) Radiation effects on crystalline polymers – I. Gamma-radiation-induced crosslinking and structural characterization of polyethylene oxide. *International Journal of Radiation Applications and Instrumentation. Part C. Radiation Physics and Chemistry, 40* (6), 501–505.

[4] Jha, A.K., Hule, R.A., Jiao, T., et al. (2009) Structural analysis and mechanical characterization of hyaluronic acid-based doubly cross-linked networks. *Macromolecules, 42* (2), 537–546.

[5] Enikopolov, N.S. (1991) Solid phase chemical reactions and new technologies. *Russian Chemical Reviews, 60* (3), 283–287.

[6] Khabarov, V.N., Zelenetskii, A.N. (2008) Nanotechnological reticulation of hyaluronic acid (in Russian). *Kosmetik International, 2*, 8–15. Хабаров В.Н., Зеленецкий А.Н. (2008) Нанотехнологическая ретикуляция гиалуроновой кислоты. *Kosmetik International, 2*, 8–15.

[7] Khabarov, V.N., Selyanin M.A., Zelenetskii A.N. (2008) Solid-state modification of hyaluronic acid for applications in esthetic medicine (in Russian), *Vestnik Estetichskoi Mediciny, 7* (3), 18–24.

Хабаров, В.Н., Селянин, М.А., Зеленецкий, А.Н. (2008) Твердотельная модификация гиалуроновой кислоты для целей эстетической медицины, *Вестник эстетической медицины,* **7** (3), 18–24.

[8] Prut, E.V., Zelenetskii, A.N. (2001) Chemical modification and blending of polymers in an extruder reactor. *Russian Chemical Reviews,* **70** (1) 65–80.

[9] Volkov, V.P., Zelenetskii, A.N., Akopova, T.A. et al. (2007) The method of obtaining cross-linked hyaluronic acid salts. Russian Federation Patent 2366665, filed Dec. 03, 2007, and issued Oct. 09, 2009.

[10] Kablik, J., Monheit, G.D., LiPing, Y., et al. (2009) Comparative physical properties of hyaluronic acid dermal fillers. *Dermatologic Surgery,* **35**, 302–312.

[11] Kuo, J.W. (2006) *Practical Aspects of Hyaluronan Based Medical Products.* Taylor & Francis Group, New York.

[12] Semenov, N.N. (1968) *Chain Reactions* (in Russian), Nauka, Moscow Семенов, Н.Н. (1968) Цепные реакции. Наука, Москва.

[13] Garg, H.G., Hales, C.A. (eds) (2004) *Chemistry and Biology of Hyaluronan.* Elsevier, Amsterdam.

[14] Margolina, A., Ernandes, E. (2005) *New Cosmetology* (in Russian). Klavel, Moscow. Марголина, А., Эрнандес, Е. (2005) *Новая косметология.* Клавель, Москва

[15] Spicer, A.P., McDonald, J.A. (1998) Characterization and molecular evolution of a vertebrate hyaluronan synthase gene family. *Journal of Biological Chemistry,* **273**, 1923–1932.

[16] Presti, D., Scott, J.E. (1994) Hyaluronan – mediated protective effect against cell damage caused by enzymatically produced hydroxyl radical is dependent on hyaluronan molecular mass. *Cell Biochemistry & Function,* **12**, 281–288.

[17] Khabarov, V.N., Boikov, V.N., Tchizhova, N.A. et al. Molecular weight of hyaluronic acid in preparations for aesthetic medicine (in Russian). *Vestnik Estetichskoi Mediciny,* **8** (4), 16–19. Хабаров, В.Н., Бойков, П.Я., Чижова, Н.А. *и др* (2009) Значение параметра молекулярной массы гиалуроновой кислоты в препаратах для эстетической медицины. *Вестник эстетической медицины,* **8** (4), 16–19.

[18] Khabarov, V.N., Mikhailova, N.P., Selyanin, M.A. (2011) Estimation of antioxidant effeciency of biocompounds attractive for use in aesthetic medicine. *Vestnik Estetichskoi Mediciny,* **10** (1), 52–56. Хабаров, В.Н., Михайлова, Н.П., Селянин, М.А. (2011) Оценка антиоксидантной эффективности биосоединений, перспективных для применения в эстетической медицине. *Вестник эстетической медицины,* **10** (1), 52–56.

[19] Blyumenfel'd, L.A. (1977) *Problems of Biological Physics* (in Russian). Nauka, Moscow Блюменфельд, Л.А. (1977) *Проблемы биологической физики.* Наука, Москва

[20] Blyumenfel'd, L.A., Kol'tover, V.K. (1972) Energy transformation and conformational transitions in mitochondrial memebranes as relaxation processes (in Russian). *Molekuliarnaia Biologiia,* **6**, 161–166. Блюменфельд, Л.А., Кольтовер, В.К. (1972) Трансформация энергии и конформационные переходы в митохондриальных мембранах как релаксационные процессы. *Молекулярная биология,* **6**, 161–166.

[21] Obukhova, L.K., Emanuel,' N.M. (1983) The role of free-radical oxidation reactions in the molecular mechanisms of the ageing of living organisms. *Russian Chemical Reviews,* **52** (3), 201–212.

[22] Pryor, W.A. (1976) The role of free radical reactions in biological systems, in *Free Radicals in Biology,* vol. 1 (ed. W.A. Pryor), Academic Press Inc., New York, pp. 1–50.

[23] Fridovich, Y. (1997) Superoxide anion radical, syperoxide dismutases, and related matters *Journal of Biological Chemistry,* **272**, 18515–18517.

[24] Osipov, A.N., Azizova, O.A., Vladimirov, Y.V. (1990) Active forms of oxygen and their role in organisms (in Russian). *Uspekhi Biologicheskoi Khimmii,* **31**, 180–208. Осипов, А.Н., Азизова, О.А., Владимиров, Ю.В. (1990) Активные формы кислорода и их роль в организме. *Успехи биологической химии,* **31**, 180–208.

[25] Khabarov, V.N., Selyanin, M.A., Zelenetskii, A.N. (2008) Perspectives of development of new preparations for biorevitalization (in Russian). *Vestnik Estetichskoi Mediciny,* **7** (4), 40–46. Хабаров, В.Н., Селянин, М.А., Зеленецкий, А.Н. (2008) Перспективы создания новых препаратов для биоревитализации. *Вестник эстетической медицины,* **7** (4), 40–46.

[26] Margolina, A., Ernandes, E. (2007) *New Cosmetology, Vol. 2* (in Russian). Klavel, Moscow. Марголина, А., Эрнандес, Е. (2005) *Новая косметология, Том 2*. Клавель, Москва.

[27] Sharpatyi, V.A. (2006) *Radiation Chemistry of Biopolymers*. CRC Press, Taylor & Francis Group, Boca Raton.

[28] Khabarov, V.N., Kozlov, L.L., Panchenkov, G.M. (1980) Gamma-radiolysis of aqueous solutions of methyl orange and chrysoidine (in Russian). *Khimiya Vysokikh Energii* **14** (5), 406–408. Хабаров, В.Н., Козлов, Л.Л., Панченков, Г.М. (1980) Гамма-радиолиз водных растворов метилоранжа и хризоидина. *Химия высоких энергий,* **14** (5), 406–410.

[29] Khabarov, V.N., Kozlov, L.L., Panchenkov, G.M. (1981) Mechanism and kinetics of discoloration of azo dyes in aqueous solutions upon gamma-irradiation (in Russian). *Khimiya Vysokikh Energii,* **15** (3), 218–222. Хабаров, В.Н., Козлов, Л.Л., Панченков, Г.М. (1981) Механизм и кинетика обесцвечивания азокрасителей в водных растворах при действии гамма-излучения, *Химия высоких энергий,* **15** (3), 218–222.

[30] Khabarov, V.N., Kozlov, L.L., Panchenkov, G.M. (1981) Kinetic dependences of radiation-chemical reduction of azo dyes in aqueous solutions (in Russian) *Zhurnal Fizicheskoi Khimii,* **55** (12), 3072–3075. Хабаров, В.Н., Козлов, Л.Л., Панченков, Г.М. (1981) Кинетические закономерности радиационно-химического восстановления азокрасителей в водных растворах и полимерных пленках. *Журнал физической химии,* **55** (12), 3072–3077.

[31] Khabarov, V.N., Kozlov, L.L., Panchenkov, G.M. (1981) Study on radiation-chemical transformations of polyamide (in Russian). *Khimiya Vysokikh Energii,* **15** (6), 487–491. Хабаров В.Н., Козлов Л.Л., Панченков Г.М. (1981) Исследования радиационно-химических превращений полиамида. *Химия высоких энергий.* **15** (6), 487–491.

[32] Khabarov, V.N., Kozlov, L.L., Panchenkov, G.M. (1982) Radiation-induced oxidation of polyamide (in Russian). *Plastmassy,* **6**, 29–32. Хабаров, В.Н., Козлов, Л.Л., Панченков, Г.М. (1982) Радиационное окисление полиамида, *Пластмассы,* **6**, 29–32.

[33] Khabarov, V.N., Kozlov, L.L., Panchenkov, G.M. (1983) Spectroscopy study of radiation-chemical transformations in polyamide (in Russian) *Zhournal Prikladnoi Spektroskopii,* **39** (3), 444–449. Хабаров, В.Н., Козлов, Л.Л., Панченков, Г.М. (1983) Спектроскопические исследования радиационно-химических превращений полиамида. *Журнал Прикладной спектроскопии,* **39** (3), 444–449.

[34] Volkov, V.P., Zelenetskii, A.N., Khabarov, V.N., Selyanin, M.A. (2008) The method of obtaining salts of modified hyaluronic acid cross-linked with ascorbic acid and bioactive composition thereof. Russian Federation Patent No. 2382050, filed Jun. 05, 2008 and issued Feb. 20, 2010.

[35] Volkov, V.P., Zelenetskii, A.N., Khabarov, V.N., Selyanin, M.A. (2008) The method of obtaining salts of modified hyaluronic acid cross-linked with tocopherol. Russian Federation Patent No. 2382052, filed Jul. 09, 2008, and issued Feb. 20, 2010.

[36] Volkov, V.P., Zelenetskii, A.N., Khabarov, V.N., Selyanin, M.A. (2008) The method of obtaining salts of cross-linked hyaluronic acid modified with vitamins. Russian Federation Patent No. 2387671, filed Jul. 30, 2008, and issued Apr. 27, 2010.

[37] Volkov, V.P., Zelenetskii, A.N., Khabarov, V.N., Selyanin, M.A. (2008) The method of obtaining salts of cross-linked hyaluronic acid modified with folic acid. Russian Federation Patent No. 2387670, filed Jul. 30, 2008, and issued Apr. 27, 2010.

[38] Volkov, V.P., Zelenetskii, A.N., Khabarov, V.N., Selyanin, M.A. (2008) The method of obtaining salts of cross-linked hyaluronic acid modified with retinol. Russian Federation Patent No. 2386641, filed Jul. 09, 2008, and issued Jan. 20, 2010.

[39] Volkov, V.P., Zelenetskii, A.N., Khabarov, V.N., Selyanin, M.A. (2008) The method of obtaining salts of cross-linked hyaluronic acid modified with riboflavin. Russian Federation Patent No. 2386640, filled Jun. 25, 2008, and issued Apr. 20, 2010.

[40] Khabarov, V.N., Selyanin, M.A., Mikhailova, N.P., Zelenetsky, A.N. (2009) Bioactive compositions comprising hyaluronic acid modified in solid-phase (in Russian). *Vestnik Estetichskoi Mediciny,* **8** (1), 49–53. Хабаров, В.Н., Селянин, М.А., Михайлова, Н.П., Зеленецкий, А.Н. (2009) Биоактивные композиции на основе модифицированной гиалуроновой кислоты. *Вестник эстетической медицины,* **8** (1), 49–54.

[41] Khabarov, V.N., Mikhailova, N.P., Selyanin, M.A. (2010) Biorepairs – a new class of injectable preparations (in Russian). *Esteticheskaya Medicina,* **9** (2), 3–12. Хабаров, В.Н., Михайлова,

Н.П., Селянин, М.А. (2010) Биорепаранты- новый класс инъекционных препаратов. *Эстетическая медицина,* **9** (2), 3–12.

[42] Khabarov, V.N., Selyanin, M.A., Mikhailova, N.P. (2010) Innovative technologies in injection cosmetology from viewpoint of bioorganic chemistry: 'Hyalrepair' line of products (in Russian) *Vestnik Estetichskoi Mediciny,* **9** (2), 69–76. Хабаров, В.Н., Селянин, М.А., Михайлова, Н.П. (2010) Инновационные технологии в инъекционной косметологии с точки зрения биоорганической химии: линия «Гиалрипайер». *Вестник эстетической медицины,* **9** (2), 69–76.

[43] Petrukhina, A.O. (2004) *UV-Irradiation and Skin: Effects, Problems, Solutions* (in Russian). Nauka Medicina Veterinaria, Moscow. bПетрухина, А.О. (ред) (2004) *УФ-излучение и кожа: эффекты, проблемы, решения* Наука Медицина Ветеринария, Москва.

[44] Millis, A. (1992) Differential expression of metalloproteinase and tissue inhibitor of metalloproteinase genes in aged human fibroblast. *Experimental Cell Research,* **201**, 373–378.

[45] Strand, F.L. (1999) *Neuropeptides: Regulators of Physiological Processes.* The MIT Press, Cambridge.

[46] Ivanova, L.N., Babina, A.V., Baturina, G.S., Katkova, L.E. (2013) Effect of vasopressin on the expression of genes for key enzymes of hyaluronan turnover in Wistar Albino Glaxo and Brattleboro rat kidneys. *Experimental Physiology,* **98** (11), 1608–1619.

[47] Fricke, R., Hartmann, F. (1974) *Connective Tissue: Biochemistry and Pathophysiology.* Springer, Berlin.

[48] Burlakova, E.B., Grechenko, T.N., Sokolov, E.N., Terekhova, S.F. (1986) Effect of inhibitors of radical reactions of lipid oxidation on the electrical activity of snail isolated neurons (in Russian). *Biofizika,* **31** (5), 921–923.

[49] Burlakova, E.B. (1994) Effect of ultra small doses (in Russian). *Vestnik Rossiiskoi Akademii Nauk,* **64**, (5), 425–434.

[50] Burlakova, E.B., Boĭkov, P.Y., Papina, R.I., Kartsev, V.G. (1996) Bimodal effect of picolinic acid derivatives on the rate of plant cells proliferation (in Russian). *Izvestiya Rossiiskoi Akademii Nauk – Seriya Biologicheskaya,* **1**, 30–35.

[51] Konovalikhin, S.V, Boĭkov, P.Y., Burlakova, E.B. (2000) Structure-physiological activity correlation for picolinic acid derivatives based on quantum chemical data (in Russian). *Izvestiya Rossiiskoi Akademii Nauk – Seriya Biologicheskaya,* **2**, 153–157.

[52] Balazs, E.A., Watson, D., Duff, J.F., Roseman S. (1967) Hyaluronic acid in synovial fluid. 1. Molecular parameters of hyaluronic acid in normal and arthritis human fluids. *Arthritis & Rheumatism,* **10**, 357–376.

[53] Matveeva, E.L. (2007) Study of age-related changes and gender features of the biochemical composition of human knee joint synovial fluid (in Russian). *Klinicheskaya Laboratornaia Diagnostika,* **5**, 15–17.

[54] Vo, N., Niedernhofer, L.J., Nasto, L.A. et al. (2013) An overview of underlying causes and animal models for the study of age-related degenerative disorders of the spine and synovial joints. *Journal of Orthopaedic Research,* **31**(6), 831–837.

[55] Lee, A.S., Ellman, M.B., Yan, D. (2013) A current review of molecular mechanisms regarding osteoarthritis and pain. *Gene,* **527** (2), 440–447.

[56] Vallières, M., du Souich, P. (2010) Modulation of inflammation by chondroitin sulfate. *Osteoarthritis Cartilage, Suppl.* **1**, S1–S6.

[57] Monfort, J., Pelletier, J.P., Garcia-Giralt, N., Martel-Pelletier, J. (2008) Biochemical basis of the effect of chondroitin sulphate on osteoarthritis articular tissues. *Annals of Rheumatic Diseases,* **67** (6), 735–740.

[58] Im, G.I., Choi, Y.J. (2013) Epigenetics in osteoarthritis and its implication for future therapeutics. *Expert Opinion on Biological Therapy,* **13** (5), 713–721.

[59] Goldberg, V.M., Buckwalter, J.A. (2005) Hyaluronans in the treatment of osteoarthritis of the knee: evidence for disease-modifying activity. *Osteoarthritis Cartilage,* **13** (3), 216–224.

[60] Dougados, M. (2000) Sodium hyaluronan therapy in osteoarthritis: Arguments for a potential beneficial structural effect. *Seminars in Arthritis and Rheumatism,* **30** (2), 19–25.

[61] Radaeva, I.F., Kostina G.A. (1998) Use of hyaluronic acid for the treatment of various pathologic states. *Pharmaceutical Chemistry Journal,* **32** (9), 492–494.

[62] Tobetto, K., Yasui, T., Ando, T. *et al* (1992) Inhibitory effects of hyaluronan on ^{14}C –arachidonic acid release from labeled human synovial fibroblasts. *Japanese Journal of Pharmacology,* **60**, 79–84.

[63] Pavlova, V.A. (1980) *Synovial Medium of the Joints* (in Russian). Medicina, Moscow. Павлова, В.А. (1980) *Синовиальная среда суставов,* Медицина. Москва

[64] Belen'kii, (2005) *Preparations of Hyaluronic Acid in Osteoarthritis Treatment* (in Russian). Medicina, Moscow.

[65] Moscowitz, R.W., Altman R.D., Hochberg, M.C., et al. (2007) *Osteoarthritis: Diagnosis and Medical/Surgical Management.* Lippincott Williams & Wilkins, Philadelphia.

[66] Balazs, E.A. (2003) Analgesic effect of elastoviscous hyaluronan solutions and the treatment of arthritic pain. *Cell Tissues Organs,* **174** (1–2), 49–62.

[67] Laurent, T.C., Fraser, J.R. (1992) Hyaluronan. *Journal of Federation of American Societies of Experimental Biology,* **6** (7), 2397–23404.

[68] Wang, C.-T, Lin, Y.-T., Chiang, B.-L., et al. (2006) High molecular weight hyaluronic acid down-regulates the gene expression of osteoarthritis-associated cytokines and enzymes in fibroblast-like synoviocytes from patients with early osteoarthritis. *Osteoarthritis and Cartilage,* **14**, 1237–1242.

[69] Nagano, O., Saya, H. (2004) Mechanism and biological significance of CD44 cleavage. *Cancer Science,* **95** (12), 930–935.

[70] Knudson, C.B., Knudson, W. (2004) Hyaluronan and CD44: modulators of chondrocyte metabolism. *Clinical Orthopaedics and Related Research,* (**427 Suppl**), 152–162.

[71] Hall, B.K. (2005) *Bones and Cartilage: Developmental and Evolutionary Skeletal Biology.* Elsevier Academic Press, San Diego

[72] von der Mark, K., Gauss, V., von der Mark, H., Müller, P. (1977) Relationship between cell shape and type of collagen synthesised as chondrocytes lose their cartilage phenotype in culture. *Nature,* **267**, 531–553

[73] Nikolaeva, S.S. Chkhol, K.Z., Bykov, V.A. et al. (2002) Biochemical characteristics and water exchange in the surface layer of human joint cartilage. *Bulletin of Experimental Biology and Medicine,* **134** (4), 335–337.

[74] Zavadovskii, B.V., Kovalenko, E.A. (1999) Connection of the level of antibodies to the glycosaminoglycans of cartilage in the patients with osteoarthrosis with the effectiveness of treatment by chondroprotective agents (in Russian). *Terapevticheskii Arkhiv,* **5**, 45–48. Завадовский, Б.В., Коваленко, Е.А. (1999) Связь уровня антител к гликозаминогликанам хряща у больных остеоартрозом с эффективностью лечения хондропротекторами. *Терапевтический архив,* **5**, 45–48.

[75] Goa, K.L., Benfield, P. (1994) Hyaluronic acid. A review of its pharmacology and use as a surgical aid in ophthalmology and its therapeutic potential in joint disease and wound healing. *Drugs,* **47** (3), 536–560.

[76] Peck, C.M., Joos, Z.P., Zaugg, B.E. et al. (2009) Comparison of the corneal endothelial protective effects of Healon-D and Viscoat. *Clinical & Experimental Ophthalmology,* **37** (4), 397–401.

[77] Austa, L., Clark, C., Turle, E.A., Vandelig, K. (1991) Hyaluronan and cell-associated hyaluronan binding protein regulate the locomotion of ras-transformed cell. *Journal of Cellular Biology,* **112** (5), 1041–1047.

[78] Isacke, C.M., Yarwood, H. (2002) The hyaluronan receptor, CD44. *The International Journal of Biochemistry & Cell Biology,* **34** (7), 718–721.

[79] Ponta, H., Sherman, L., Herrlich, P.A. (2003) CD44: From adhesion molecules to signalling regulators. *Nature Reviews Molecular Cell Biology* **4**, 33–45.

[80] Rooney, P., Kumar, S., Ponting, J., Wang, M. (1995) The role of hyaluronan in tumor neovascularization. *International Journal of Cancer,* **60**, 632–636.

[81] Delpech, B., Girard, N., Bertrand, P., Courel, M.N. (1997) Hyaluronan: fundamental principles and applications in cancer. *Journal of Internal Medicine,* **7**, 41–48.

[82] West, D.C. (1985) Angiogenesis induced by degradation products of hyaluronic acid. *Science,* **228**, 1324–1326.

[83] Filion, M.C., Menard, S., Filion, B., et al. (2002) Anti-cancer activity of hyaluronan, in *Hyaluronan: Proceedings of an International Meeting, September 2000, North East Wales Institute, UK* (eds. J.F. Kennedy, G.O. Phillips, P.A. Williams, V.C. Hascall), Woodhead Publishing Ltd.=, Cambridge, pp. 419–427.

[84] Shevchenko, N.A., Pavlova, V.M., Boĭkov, P.Y. et al. (1986) Correlation of the activity of the key enzymes of glycolysis and respiration in hepatomas with different growth rates (in Russian). *Eksperimental'naia Onkologiia,* **8** (3), 40–43. Шевченко, Н.А., Павлова, В.М., Бойков, П.Я., Богданов, Г.Н. (1986) Соотношение активности ключевых ферментов гликолиза и дыхания в гепатомах с различной скоростью роста. *Экспериментальная онкология,* **8** (2), 40–43.

[85] Coon, H.G. (1966) Clonal stability and phenotypic expression of chick cartilage cells in vitro. *Proceedings of the National Academy of Sciences of the United States of America,* **55**, 66–73.

[86] Boĭkov, P.Y. (1986) *Mechanisms of Proliferation Initiation in Differentiated Cells.* PhD Thesis. Moscow State University. bБойков, П.Я. (1986) *Механизмы инициирования пролиферации дифференцированных клеток.* Диссертация на соискание ученой степени доктора биологических наук. Москва.

[87] Evanko, S.P., Wight, T.N. (1999) Intracellular localization of hyaluronan in proliferating cells. *The Journal of Histochemistry and Cytochemistry,* **47** (10), 1331–1342.

[88] Xu, H., Ito. T., Tawada. A. (2002) Effect of hyaluronan oligosaccharides on the expression of heat shock protein 72. *Journal of Biological Chemistry,* **277**, 17308– 17314.

[89] Louderbough, J., Schroeder, J. (2011) Understanding the dual nature of CD44 in breast cancer progression. *Molecular Cancer Research,* **9** (12), 1573–1586.

[90] Okamoto, J. (2001) Proteolitic release of CD44 intracellular domain and its role in the CD44 signaling pathway. *Journal of Cellular Biology,* **155**, 755–762.

[91] Shkil,' N.N., Glotov, A.G., Glotova, T.I. (2003) Influence of hyaluronic acid gel on interferonogenesis in mice (in Russian). *Voprosy Virusologii,* **48** (5), 26–29.

[92] Barinskii, I.F., Alimbarova, L.M., Samoilenko, I.I. (1988) Development of suppository form of interferon inductor 'poludan' and potentiation of its activity by hyaluronic acid (in Russian). *Voprosy Virusologii,* **10**, 237–239. Баринский, И.Ф., Алимбарова, Л.М., Самойленко, И.И. (1998) Разработка свечевой формы индуктора интерферона полудана и потецирование его действия гиалуроновой кислотой. *Вопросы вирусологии,* **10**, 237–239.

[93] Pure, E., Culf, C.A. (2001) A crucial role for CD44 in inflammation. *Trends in Molecular Medicine,* **7**, 213–221.

[94] Siegelman, M.N., DeGrendelle, H.C., Estess, P. (1999) Activation and interaction of CD44 and hyaluronan in immunological system. *Journal of Leukocyte Biology,* **66** (2), 315–321.

[95] Lesley. J., Hyman. R. (1998) CD44 structure and function. *Frontiers in Bioscience,* **3**, 616–630.

[96] Bajorath. J. (2000) Molecular organisation, structural features, and ligand binding characteristic of CD44, a highly variable cell surface glycoprotein with multiple functions. *Proteins,* **39**, 103–111.

[97] Ibragimov, R.M., Khasanov, A.G., Kayumov, F.A., Sufiyarov I.F. (2010) Prophylaxis of incompetence of hollow organs anastomoses by hyaluronic acid-based bioexplant (in Russian). *Rossiiskie Medicinskie Vesti,* **15** (2), 45–50. Ибрагимов, Р.М., Хасанов, А.Г., Каюмов, Ф.А., Суфияров, И.Ф. Профилактика несостоятельности анастомозов полых органов с помощью биоэксплантанта на основе модифицированной гиалуроновой кислоты. *Российские медицинские вести,* **15** (2), 45–50.

[98] Chernousov, A.F., Khorobrykh, T.V., Antonov O.N. (2005) Prophylaxis of insufficiency of gastrointestinal anastomoses (in Russian) *Khirurgiya,* **12**, 25–29. Черноусов, А.Ф., Хоробрых, Т.В., Антонов, О.Н. (2005) Профилактика недостаточности анастомозов желудочно-кишечного тракта. *Хирургия,* **12**, 25–29.

[99] Brown, T.J., Alcorn, D., Fraser, J.R. (1999) Absorbtion of hyaluronan applied to the surface of intact skin. *Journal of Investigative Dermatology,* **113** (5), 740–746.

[100] Maytin, E.V., Chung, H.H., Seetharaman, V.M. (2004) Hyaluronan participates in the epidermal response to disruption of the permeability barrier in vivo. *American Journal of Phathology,* **165** (4), 1331–1338.

[101] Dasgeb, B., Phillips, T. (200) What are scars?, in *Scar Revision,* 1st edn, (ed. K.A. Arndt), Elsevier/Saunders, Philadelphia, pp. 1–16.

[102] Stroitelev, V.V., Fedorischev, I.A. (1997) Hyaluronic acid – biologically active compound with protective and immunostimulating functions (in Russian). *Vestnik Novykh Medicinskikh Tekhnologii,* **4** (3), 98–102. Строителев, В.В., Федорищев, И.А. (1997) Гиалуроновая кислота – биологически активное вещество, обладающее защитными и иммуномодулирующими функциями. *Вестник новых медицинских технологий,* **4** (3), 98–102.

[103] Stern, R., Frost, G.I., Shuster, S. (1998) Hyaluronic acid and skin. *Cosmetics & Toiletries,* **113**, 43–48.

[104] Samoilenko, I.I., Fedorischev, I.A. (1997) Perspectives of development of highly efficient medical preparations with hyaluronic acid on the basis of the 'Hyaplus' substance. *Vestnik Novykh Medicinskikh Tekhnologii,* **3** (3), 82–83. Самойленко, И.И., Федорищев, И.А. (1996) Перспективы получения высокоэффективных лекарственных средств с гиалуроновой кислотой на базе субстанции «Гиаплюс». *Вестник новых медицинских технологий,* **3** (3), 82–83.

Conclusion

Significant information on the chemical structure, macromolecular characteristics, biological properties and medical application of hyaluronan has accumulated in international literature. The understanding of the molecular mechanisms of hyaluronic acid in the body is still limited but it is clear that it has played a crucial role in the historical development (phylogenicity) of chordates and ontogenicity of modern higher vertebrates. It is part of the main different types of intercellular substance of connective tissue, is present in large amounts in the vitreous body of the eye, synovial fluid of joints, oviducts, Wharton's jelly of the umbilical cord, skin, walls of arteries and veins, heart valves, the glomerular basement membrane of the kidneys and so on.

Since the discovery of hyaluronan, there has been a significant evolution of the views on this biopolymer. The polysaccharide was initially considered to be a passive structural component of the extracellular matrix. Now, however, it has become apparent that it is dynamically involved into many biological processes, from reproduction modulation, cell migration and differentiation during embryogenesis, to regulation of processes of inflammation and wound healing and cancer cell metastasis. Hyaluronan has multiple physiological functions in the body. It is the basis of the functioning of the mucolytic system (hyaluronan – hyaluronidase) that determines, in particular, permeability of tissue and vascular circulatory system, the resistance to penetration of infection and substance filtration in the kidneys. Hyaluronan acts as a structure that forms compounds and stores water in the gel-like matrix of the extracellular matrix. This function provides mechanical support, tissue turgor, resistance to compressive pressure and determines the damping and anti-friction properties of the synovial fluid of joints. Hyaluronan is involved in the fertilization of oocytes, as well as their division, migration, maintenance of the cells in the differentiation state and transitions of the differentiated cells into the cell cycle and back. Hyaluronan participates in the functioning of cell signalling systems.

Hyaluronic Acid: Preparation, Properties, Application in Biology and Medicine, First Edition.
Mikhail A. Selyanin, Petr Ya. Boykov and Vladimir N. Khabarov.
© 2015 John Wiley & Sons, Ltd. Published 2015 by John Wiley & Sons, Ltd.

Such a wide variety of the biological properties of hyaluronan are first of all functions of molecular weight, polymorphism of the different structural forms, physicochemical properties of molecules of different molecular weight and the ionic environment and concentration of the biopolymer in tissues and organs. Upon cleavage of polysaccharide macromolecules, the different polymorphous structures with different structures and functional properties are formed. The formed fragments have broad and often contradictory functions and properties. Short biopolymer chains often behave as 'alarm signals' whereas longer fragments act as triggers for the signalling systems. For example, tetrasaccharides induce heat shock proteins that are supposed to heal the cell damage and inhibit apoptosis, leading to cell death. Similar biological properties of hyaluronan degradation products, which help cell preservation, represent the brilliant example of extending Le Chatelier's principle on biological systems. Such properties confirm the validity of the concept that the degradation products of biopolymers promote restoration and/or preservation of system homeostasis and can be used as control factors of the regenerative process in the environment in which they are present. Hence, this is a natural way, by which a strategy for medical use of known products based on hyaluronan could be developed and the new generation of the therapeutic products could be created.

In this book we have tried to consider many of issues with different degrees of complexity and completeness of the presented material. It seems that by choosing one particular problem, for example, the biological role of hyaluronan, we could protect the reader from the inexhaustible variety of other no less interesting but nevertheless different areas of knowledge and data. However, we intentionally used a multidisciplinary approach while being aware that some issues are described superficially and require address to additional special literature.

We realize that this book presents more questions than answers. It should serve only as an introduction to the infinitely interesting and still largely unknown world of this truly amazing macromolecule faced by biologists, physicians, physicists and chemists. Therefore, we would like to finish the book with the words of the famous Israeli professor, Shlomo Sand:

> ... it is a known fact: to stay outside a specific sphere of the study and balance on the borders that divide different spheres can sometimes allow for the appearance of a non-standard view of things and reveal unexpected links between them. It is quite often thinking 'outside' rather than 'inside' that can enrich the scientific mindset, in spite of all the weaknesses associated with a lack of speciality ...

Index

Hyaluronic Acid: Preparation, Properties, Application in Biology and Medicine, First Edition.
Mikhail A. Selyanin, Petr Ya. Boykov and Vladimir N. Khabarov.
© 2015 John Wiley & Sons, Ltd. Published 2015 by John Wiley & Sons, Ltd.